APPLIED MATHEMATICS FOR ECONOMICS

THEORY AND DECISION LIBRARY

General Editors: W. Leinfellner and G. Eberlein

Series A: Philosophy and Methodology of the Social Sciences
Editors: W. Leinfellner (Technical University of Vienna)
G. Eberlein (Technical University of Munich)

Series B: Mathematical and Statistical Methods
Editor: H. Skala (University of Paderborn)

Series C: Game Theory, Mathematical Programming and Mathematical Economics
Editor: S. Tijs (University of Nijmegen)

Series D: System Theory, Knowledge Engineering and Problem Solving
Editor: W. Janko (University of Vienna)

SERIES B: MATHEMATICAL AND STATISTICAL METHODS

Editor: H. Skala (Paderborn)

Editorial Board

J. Aczel (Waterloo), G. Bamberg (Augsburg), W. Eichhorn (Karlsruhe),
P. Fishburn (New Jersey), D. Fraser (Toronto), B. Fuchssteiner (Paderborn),
W. Janko (Vienna), P. de Jong (Vancouver), M. Machina (San Diego),
A. Rapoport (Toronto), M. Richter (Karlsruhe), D. Sprott (Waterloo),
P. Suppes (Stanford), H. Theil (Florida), E. Trillas (Madrid), L. Zadeh (Berkeley).

Scope

The series focuses on the application of methods and ideas of logic, mathematics and statistics to the social sciences. In particular, formal treatment of social phenomena, the analysis of decision making, information theory and problems of inference will be central themes of this part of the library. Besides theoretical results, empirical investigations and the testing of theoretical models of real world problems will be subjects of interest. In addition to emphasizing interdisciplinary communication, the series will seek to support the rapid dissemination of recent results.

JATI K. SENGUPTA

Department of Economics,
University of California, Santa Barbara, U.S.A.

APPLIED MATHEMATICS FOR ECONOMICS

D. REIDEL PUBLISHING COMPANY

A MEMBER OF THE KLUWER ACADEMIC PUBLISHERS GROUP

DORDRECHT / BOSTON / LANCASTER / TOKYO

Library of Congress Cataloging in Publication Data

Sengupta, Jati Kumar.
  Applied mathematics for economics.

  (Theory and decision library. Series B, Mathematical and Statistical methods).
  Bibliography: p.
  Includes index.
  1. Economics—Mathematics. I. Title. II. Series.
HB71.S398 1987   330′.028   87-20544
ISBN 90-277-2588-8

Published by D. Reidel Publishing Company,
P.O. Box 17, 3300 AA Dordrecht, Holland.

Sold and distributed in the U.S.A. and Canada
by Kluwer Academic Publishers,
101 Philip Drive, Assinippi Park, Norwell, MA 02061, U.S.A.

In all other countries, sold and distributed
by Kluwer Academic Publishers Group,
P.O. Box 322, 3300 AH Dordrecht, Holland.

All Rights Reserved
© 1987 by D. Reidel Publishing Company, Dordrecht, Holland
No part of the material protected by this copyright notice may be reproduced or
utilized in any form or by any means, electronic or mechanical
including photocopying, recording or by any information storage and
retrieval system, without written permission from the copyright owner

Printed in The Netherlands

Preface

Modern economics has increasingly used mathematical concepts and methods. Tools of applied mathematics have therefore become all the more important today to a modern student of economics. This text provides an intermediate level mathematical introduction to economic theory. It emphasizes only those applied mathematical tools which are in my opinion most useful to the understanding of the basic economic models used in modern economics today. It is intended to be a mathematical introduction to economics and not an economic introduction to mathematics.

Mathematical treatment is kept at a fairly elementary level, although the reader is encouraged through illustrative applications to see the economic implications of more general models involving e.g., general equilibrium theory.

The operational methods and examples used here should prove immensely helpful to applied researchers in economics and business management. Each chapter is so arranged that more difficult methods are treated at the end, so that a beginning student may skip the advanced materials in his or her first reading.

I have been fortunate in my association with many gifted students, who taught me in many ways the truth that mathematics is a language and a very universal language. Though imperfect in many ways it still provides an excellent tool for communicating economic ideas. It cannot generate economic intuition, but it can supplement and activate it.

I would never have completed this text without the constant support and encouragement from my wife. My deepest appreciation to her.

J.K.S.

Table of Contents

CHAPTER ONE		INTRODUCTION TO MATHEMATICAL ECONOMICS	1
1.1	Examples of Economic Models		2
1.2	Mathematical Preliminaries		7
	1.2.1	Scalar Number Systems	7
	1.2.2	Functions and Equations	8
	1.2.3	Vector and Matrix Algebra: Elementary Aspects	11
	1.2.4	Random Numbers in Economics	18
1.3	Some Advanced Concepts		19
	1.3.1	Continuous Functions: Their Uses	19
	1.3.2	Taylor Series Expansion	21
	1.3.3	Hessian Matrix and Concavity	24
	1.3.4	Implicit Functions	25
1.4	Advanced Concepts in Matrix Algebra		28
	1.4.1	Rank and Linear Independence of Vectors	28
	1.4.2	Eigenvalues and Eigenvectors	32
	1.4.3	Frobenius and Metzler-Leontief Matrices	33
PROBLEMS ONE			36
CHAPTER TWO		STATIC OPTIMIZATION AND ECONOMIC THEORY	39
2.1	Linear Programming		39
	2.1.1	Simplex Algorithms	42
	2.1.2	Computational Aspects of Simplex	46
	2.1.3	Economic Applications	49
	2.1.4	Some Theory of LP	54
2.2	Nonlinear Programming		64
	2.2.1	Kuhn-Tucker Theory	64
	2.2.2	Economic Applications	72
	2.2.3	Set-theoretic Concepts of Efficiency	75
2.3	Nonlinear Maximization with Equality Constraints		79
2.4	A Critique of Optimization in Economic Theory		82
PROBLEMS TWO			83
CHAPTER THREE		DYNAMIC SYSTEMS IN ECONOMICS	86
3.1	Linear Difference Equation Systems		88
	3.1.1	First Order Equation	89
	3.1.2	Second Order Equation	92
	3.1.3	n-th Order Equation	97
	3.1.4	Time-Dependent Nonhomogeneous Term	98

3.2	Linear Differential Equation Systems		104
	3.2.1	Second Order Equation	105
	3.2.2	n-th Order Equation	110
	3.2.3	Simultaneous Equations	111
3.3	Economic Applications		116
	3.3.1	Market Model with Price Expectations	116
	3.3.2	Cournot Game with Learning	119
	3.3.3	Model of Maximum Sustainable Yield	121
	3.3.4	Keynesian Macro-dynamic Model	123
3.4	Nonlinear Dynamic Equations		126
	3.4.1	Logistic Model of Growth	127
	3.4.2	Neoclassical Growth Model	128
	3.4.3	Lyapunov Theorem on Stability: Walrasian System	130
	3.4.4	Mixed Difference-Differential Equations	133
PROBLEMS THREE			136

CHAPTER FOUR	DYNAMIC OPTIMIZATION AND CONTROL		140
4.1	Intertemporal Optimization: Discrete Time		141
	4.1.1	Competitive Model with Inventories	141
	4.1.2	Stabilization in Commodity Markets	145
	4.1.3	Optimal Growth in Neoclassical Models	146
	4.1.4	Discrete Pontryagin Principle	148
4.2	Intertemporal Optimization: Continuous Time		153
	4.2.1	Optimal Model of Advertisement	156
	4.2.2	Continuous Pontryagin Principle	158
	4.2.3	Economic Applications of Optimal Control	163
	4.2.4	Dynamic Programming: Applications	171
PROBLEMS FOUR			179

CHAPTER FIVE	STATIC AND DYNAMIC GAMES		182
5.1	Forms of Games		182
	5.1.1	Extensive Form	182
	5.1.2	Normal Form	185
5.2	Two-Person Zero-Sum Games		187
	5.2.1	Game Against Nature	187
	5.2.2	Cournot Market Games	192
	5.2.3	Bi-Matrix Non-cooperative Game	200
	5.2.4.	Differential Games Over Time	204
5.3	Concepts in n-person Games		207
	5.3.1	Coalitions	207

	5.3.2 Core	208
	5.3.3 Characteristic Function	209
5.4.	Economic Applications of Core	210
PROBLEMS FIVE		216

CHAPTER SIX   SOME RECENT APPLIED ECONOMIC MODELS 220
   6.1 Stochastic Control Models 220
   6.2 Models of Rational Expectations 226
   6.3 Disequilibrium Systems 232

SELECTED READING LIST 236
INDEX 237

CHAPTER ONE
INTRODUCTION TO MATHEMATICAL ECONOMICS

Over the course of years applied mathematical economics has developed in three phases: understanding of the concept of economic equilibrium through equations and algebraic systems, analysis of optimization and stability in relation to economic theory and lastly, problems of modeling large-scale economic systems. The last phase is still continuing and includes such topics as: econometric modeling, disequilibrium analysis, and modeling economic policy as stochastic control and differential games. Besides, there are abstract constructions like purely exchange economies, super game models, and abstract general equilibrium models involving random elements.

We emphasize in this book only the operational and applied aspects of mathematical treatment arising in the first two phases, although some examples will be provided of recent developments in the third phase. We refer to some of the basic theorems in mathematical economics without explicit rigor or proof, in order to emphasize the economic insight provided by such theorems.

The following theorems are used more often in applied studies: (a) Kuhn-Tucker theorem, (b) theorems used in Leontief's input-output models, (c) minimax theorem of game theory, (d) Pontryagin's maximum principle and (e) Lyapunov's stability theorems for dynamic systems.

Mathematical economics provides an understanding of economic theory and its logical framework in three distinct ways: by decomposing its various parts, by generalizing to many variables and by checking its consistency. Thus, a macroeconomic model for the whole economy may have a microeconomic basis and the linkage between the micro and macro is a subject for mathematical economics. This is called the aggregation problem and it has very important implications for econometric models which apply mathematical economic models over observed empirical data. For the second example, consider a competitive market model for a single good with one demand and one supply function such that demand equals supply in equilibrium. What would be the formulation if we generalize this to n goods and n interrelated markets? Under what conditions there would be simultaneous equilibrium in all the markets? This question is treated in some detail in mathematical economics. As a third example consider a duopoly market game for a single good supplied by two competitors. Each competitor maximizes his profit treating other's output as given. Under what conditions the profit maximizing output strategies of the two duopolists would be mutually consistent? This is analyzed in the mathematical treatment of the

Cournot-type duopoly games.

While mathematics has helped considerably to clarify the logical and analytical basis of economic theory, it has also helped in abstract conceptualization. For example, mathematical abstractions have led to artificial conceptualizations of the real world economies such as replicated and simulated economies, hypothetical auctioneers and such game-theoretic constructs as metagames and supergames.

1.1 Examples of Economic Models

Economic models usually specify the decision problem of an economic agent or a group of agents such as the consumer or the producer and it is most convenient to view the economic model in three parts: an objective function, a set of relations between variables and some constraints on the system. Thus, a single producer in a competitive market may set his output so as to maximize profits under the constraints imposed by his production process and possibly his budget. Profit is the goal, the output as a function of the various inputs specifies the relations and the requirement that output and inputs have to be nonnegative (or positive) are the constraints on his decision. Consider some examples.

Ex 1. A producer has to decide on the amounts of two inputs, capital (y_1) and labor (y_2) to use in producing one output (x). The input-output relation is given by the function $x = f(y_1, y_2)$ and his goal is to maximize profits $z = px - q_1 y_1 - q_2 y_2$, where p, q_1, q_2 are the prices of output and the two inputs respectively. We assume that the producer is a price-taker in the sense that p, q_1, q_2 are given to him from outside. These will be called 'parameters' and denoted by θ; note that there will also be parameters in the relation $x = f(y_1, y_2)$, also called the production function which are determined by the technology of production. The variables (x, y_1, y_2) other than the parameters can be classified into two ways: (1) dependent and independent variables e.g., y_1, y_2 are independent and output x is a dependent variable. Independent variables are also called exogenous and the dependent variables endogenous, (2) state and decision variables e.g., y_1, y_2 are decision variables or control variables and x is the state variable. Once the decision variables are chosen in a specified level, the state variables are determined by the function or the equation $x = f(y_1, y_2)$. If the decision variables are chosen by maximizing the objective function such as profits, we obtain a <u>normative</u> economic model. Thus we have the following normative model

$$\begin{aligned} & \text{maximize } z = px - q_1 y_1 - q_2 y_2 \\ & (y_1, y_2) \\ & \qquad \text{subject to } x = f(y_1, y_2). \end{aligned} \tag{1.1}$$

By substituting for x in the profit function and applying the first order condition for maximization according to calculus, we obtain

$$p \frac{\partial f}{\partial y_1} = q_1; \quad p \frac{\partial f}{\partial y_2} = q_2 .$$

The production function $x = f(y_1, y_2)$ must be nonlinear in this case and also satisfy the second-order conditions for maximum profits.

<u>Ex 2</u>. A farmer produces two crops in quantities x_1(corn) and x_2(wheat) with expected net returns c_1 and c_2 per unit of output. The limits of availability of three resources are land (b_1 acres), labor (b_2 manhours) and capital (b_3 machine hours). The profit maximizing outputs are then determined by a linear programming (LP) model as follows:

$$\underset{(x_1,x_2)}{\text{maximize}} \quad z = \sum_{j=1}^{2} c_j x_j$$

subject to \hfill (1.2)

$$a_{11} x_1 + a_{12} x_2 \leq b_1 \text{ (land)}$$
$$a_{21} x_1 + a_{22} x_2 \leq b_2 \text{ (labor)}$$
$$a_{31} x_1 + a_{32} x_2 \leq b_3 \text{ (capital)}$$
$$x_1, x_2 \geq 0.$$

Here a_{ij} are the input-output coefficients for input type i and output type j e.g., a_{12} is land wheat ratio. Since the objective function and the constraints which are now inequalities are linear, the first- and second-order conditions of calculus for maximization are not applicable; some modifications are necessary and these will be discussed in later chapters in terms of Kuhn-Tucker Theorem. Note that the parameter set here is $\theta = (c_1, c_2, a_{11}, a_{12}, a_{21}, a_{22}, a_{31}, a_{32}, b_1, b_2, b_3)$ and the decision variables are (x_1, x_2). If the inputs or resources b_i were also decision variables and they could be bought in the competitive market at prices q_i (i=1,2,3), the objective function above would be modified to

$$\underset{(x_1,x_2,b_1,b_2,b_3)}{\text{maximize}} \quad z = \sum_{j=1}^{2} c_j x_j - \sum_{i=1}^{3} b_i q_i \tag{1.3}$$

and the set of decision variables would include $(x_1, x_2, b_1, b_2, b_3)$. The nonnegativity requirement on the outputs i.e., $x_1 \geq 0$, $x_2 \geq 0$ is specifically imposed in this problem. For most economic problems variables such as prices, outputs, inputs are required to be nonnegative or even positive. The economic implications of such assumptions are very important for real world applications.

Optimization in economic models, as illustrated by the two static examples above has three fundamental implications for economic theory. The first is concerned with the existence of the optimal solution if any and how to compute it. Usually if it exists, we need to characterize it by necessary and sufficient conditions. The second deals with stability or otherwise of the optimal solution in respect of the parameters. Thus if x^* is optimal for a given set θ of parameters, would it remain optimal for a slight perturbation of θ? If the neighborhood of θ around which x^* retains optimality is not very small, one might call such an optimal solution stable at least in this neighborhood. The third implication of optimality is that one optimization problem may generate a related optimization problem, also called a dual problem with a new set of variables which have interesting economic interpretations. Thus in the LP model (1.2), it can be shown that the dual problem for the farmer is one of minimizing the cost function

$$\begin{array}{ll} \underset{(\lambda_1,\lambda_2)}{\text{minimize}} & C = b_1\lambda_1 + b_2\lambda_2 + b_3\lambda_3 \\ & \text{subject to} \\ & a_{11}\lambda_1 + a_{21}\lambda_2 + a_{31}\lambda_3 \geq c_1 \\ & a_{12}\lambda_1 + a_{22}\lambda_2 + a_{32}\lambda_3 \geq c_2 \\ & \lambda_1 \geq 0,\ \lambda_2 \geq 0,\ \lambda_3 \geq 0. \end{array} \quad (1.4)$$

By the duality theorem of LP it is known that if an optimal set $x^* = (x_1^*, x_2^*)$ exists, then an optimal set $\lambda^* = (\lambda_1^*, \lambda_2^*, \lambda_3^*)$ of dual variables also exists and maximum profits equal minimal imputed costs in the sense $z^* = \Sigma\ c_j x_j^* = C^* = \Sigma\ \lambda_i^* b_i$. Any optimal dual variable λ_i^* has two economic interpretations: it is the marginal imputed cost of resource b_i at the minimal cost C^* i.e. $\lambda_i^* = \partial C^*/\partial b_i$ of resource b_i at the maximal profit z^*. In economic language λ_i^* is also called the <u>shadow price</u> of resource b_i or, its efficiency price.

The meaning of shadow price λ_i^* can be understood better if we assume that competitive input markets exist where the farmer can buy the resource b_i at a positive price q_i that is fixed for each $i=1,2,3$. Then at the maximal profit level we should have $\lambda_i^* = q_i$ for each $i=1,2,3$. Why? Because if $\lambda_i^* < q_i$ for any i, the farmer is losing for each input used at a positive level, hence he should reduce the total amount of resource b_i such that $\lambda_i^* = q_i$. For the other case if $\lambda_i^* > q_i$, he should expand his resource use due to its profit increasing effect till he reaches the point at which $\lambda_i^* = q_i$.

<u>Ex 3</u>. Consider a general equilibrium model known as the open static input-output (IO) model, where the whole economy is divided into n sectors, the output (x_i) of each sector being partly used as intermediate inputs or raw materials by other sectors (x_{ij}) and partly by households for final consumption and investment demand

(d_i):

$$x_i = \sum_{j=1}^{n} x_{ij} + d_i, \qquad i=1,2,\ldots,n.$$

If p_i denotes the price of output x_i, and w_i the cost of labor per unit of output (assuming that labor is the only final input in this model), then total return equals total cost at equilibrium for each sector:

$$p_i x_i = \sum_{k=1}^{n} p_k x_{ki} + w_i x_i \qquad i=1,2,\ldots,n.$$

If the intermediate inputs are proportional to outputs: $x_{ij} = a_{ij} x_j$ where a_{ij} are fixed positive constants (parameters) then the above two sets of equations can be written as (n=2):

$$\begin{aligned} x_1 &= a_{11} x_1 + a_{12} x_2 + d_1 \\ x_2 &= a_{21} x_1 + a_{22} x_2 + d_2 \end{aligned}$$

and

$$\begin{aligned} p_1 &= a_{11} p_1 + a_{21} p_2 + w_1 \\ p_2 &= a_{12} p_1 + p_{22} p_2 + w_2. \end{aligned}$$

Denote supply in each sector by $s_i = x_i - a_{i1} x_1 - a_{i2} x_2$ and return by $r_i = p_i - a_{1i} p_1 - a_{2i} p_2$ for $i=1,2$. Then the model assumes a two-way equilibrium for each sector: a demand-supply equilibrium $d_i = s_i$ (i=1,2) and a cost-return equilibrium $w_i = r_i$ (i=1,2). Note that this model has one final input called labor; hence if capital is another final input, it is assumed to be already included in labor (i.e. a variant of labor theory of value).

Ex 4. A monopolist sells a single product (x) in a market where the demand function has an additive error term (e):

$$p = a - bx + e \qquad p = \text{price per unit.}$$

The profit function is $z = px - (c_o + cx)$, where c_o is the fixed cost and c the marginal cost. How should the monopolist choose the optimal output x^* if his goal is to maximize profit?

For this problem the constraints imposed by the error term e are more subtle and implicit than those in Ex 2. The error term e, also called stochastic or random disturbance term is not really a single number; it is a set of numbers each occurring with a probability. For example, it may describe the following event:

value of e	-2	-1	0	+1	+2
Probability	0.3	0.2	0.0	0.2	0.3

Because of this, the demand function above is not a single demand function but a set of functions. The profit function also becomes random or stochastic, since by substitution for p we obtain $z = (a-c)x - bx^2 + ex - c_o$, where the term ex contains the random element e. The goal of maximizing profits is therefore not very well defined, since profit is not a single number (e.g. dollar values) anymore. One approach to this problem is to use expected profit as a criterion, according to which we take an average or expected value of the term (ex). This yields expected profits (Ez):

$$Ez = (a-c)x - bx^2 + (Ee)x - c_o.$$

Applying the first derivative rule for maximization one obtains the optimal output x* as

$$\frac{d(Ez)}{dx} = 0 \text{ i.e. } (a-c) - 2b\, x^* + (Ee) = 0.$$

Hence $x^* = (2b)^{-1} \{(a-c) + (Ee)\}$.

Note that if (Ee) is zero, then we get back the original deterministic optimum, where there is no stochastic disturbance term. A second approach is to replace random profits by a single number called risk adjusted profits z_A say, where the concentration or closeness of values of random profits near the average or expected value is given some weight. Thus one example of z_A is

$$z_A = Ez - w\, var(z)$$

$$= (a - c + \bar{e})\, x - (b + wv)\, x^2$$

where var(z), called the variance is a measure of the degree of closeness to the expected value and $\bar{e} = Ee$ and v is the variance measure for the random variable e i.e. $v = E\{(e - \bar{e})^2\}$; w is a nonnegative number denoting the weight on variance of profits. Again, by applying the rule $dz_A/dx = 0$, one obtains the optimal output x_A^* for risk-adjusted profits as

$$x_A^* = (a - c + \bar{e})/\{2(b + wv)\}.$$

Two points are clear. The output x_A^* is now lower than x^*, if the term (wv) is positive i.e., the higher the variance of e, the lower the optimal output. Also the price associated with x_A^* is higher than that associated with x^*. This is because price p rises as output falls.

This type of problem known as stochastic optimization in economic models occurs most frequently in recent developments of mathematical economics and econometrics. It will be discussed in later chapters.

1.2 Mathematical Preliminaries

We introduce in this section some basic mathematical concepts which are useful in analyzing economic models. The following concepts are discussed:
  (i)   Scalar number systems
  (ii) Functions and equations
  (iii) Vector and matrix algebra: elementary aspects
  (iv) Random numbers with applications

1.2.1 Scalar Number Systems

A <u>real number</u> system comprises ordinary numbers like 3, -7, 2.4, $\sqrt{2}$. This is to be distinguished from <u>complex</u> numbers which involve the imaginary quantity $\sqrt{-1}$, which is usually denoted by $i = \sqrt{-1}$. Thus 3+2i, 1.5i, -3i are complex numbers which arise in solving quadratic and higher order algebraic equations. Unlike real numbers, complex numbers cannot be ordered as higher than or lower than a different complex number. Thus 3+2i is not higher than 2+i.

A scalar is a real or complex number. This is to be distinguished from a <u>vector</u>, which is an ordered one-dimensional array of scalars i.e. a list of scalars; thus the number 5 is a scalar but (5,3) is a vector, because it is a list containing more than one scalar number.

The absolute value or, <u>modulus</u> of a real number is that number but with a positive sign; hence $|-4| = 4$, where $|\cdot|$ denotes the absolute value. For a complex number 3+4i, the absolute value is computed as
$$|3+4i| = +\sqrt{(3^2 + 4^2)} = +5.$$

For a real number x, the new variable $y = x^n$, where n is another real number is obtained by raising x to the power n: it means that x is to be multiplied by itself n times. The index n is also called the exponent of x. The following rules for the exponents are very useful: For every nonzero x and real scalars a,b,n

(i) $x^a \cdot x^b = x^{a+b}$; e.g., $3^2 \cdot 3^3 = 3^5 = 243$
(ii) $x^{-a} = 1/x^a$; e.g., $2^{-3} = 1/2^3 = 1/8$
(iii) $x^{1/n} = \sqrt[n]{x}$ = n-th root of x; e.g., $x^{1/3} = \sqrt[3]{x}$ = cube-root of x
(iv) $x^0 = x^n \cdot x^{-n} = x^n/x^n = 1$.

Economic variables like the price (p) or quantity of a good (x) are usually required to be positive or nonnegative: $p > 0$, $x > 0$ or, $p \geq 0$, $x \geq 0$. These are real scalar numbers. When we want to describe more than one good, we may need either a more general concept of a number, called a vector or, a concept of set. A set is a collection of distinct objects, where objects could be scalars or vectors. Thus, if we let S represent the set of four numbers -2, -1, 0, 1 we may write $S = \{-2,-1,0,1\}$. The set of all nonnegative numbers may be denoted by the notation $S = \{x \mid x \geq 0\}$. It reads that S contains all scalar numbers x which satisfy the given condition of nonnegativity. The vertical bar denotes 'given that'. Membership of a set is denoted by ε. Thus $1 \varepsilon S$, where $S = \{-2,-1,0,1\}$ and $\not\varepsilon$ means does not belong to e.g., $2 \not\varepsilon S$. The empty or null set is often denoted by ϕ. Just like the rules of operation for scalar numbers e.g., addition, subtraction, multiplication, etc. there are rules or laws of set operations.

1.2.2 Functions and Equations

If a scalar variable y depends on another variable x, it is called a <u>relation</u>. If the relation is such that for each x value, there exists <u>one</u> corresponding y value, then it is called a function and written as $y = f(x)$. The set of values of x is called the <u>domain</u> and the corresponding set of y values is called the <u>range</u> of the function, where x is also called the independent variable and y the dependent variable. A function may involve two or more independent variables e.g., $y = f(x,z)$ where the two independent variables are x and z. Three types of functions are most commonly used in economic models: polynomial, logarithmic and exponential. To the polynomial class belongs the following:

$$
\begin{aligned}
\text{linear:} \quad & y = a_0 + a_1 x & (n=1) \\
\text{quadratic:} \quad & y = a_0 + a_1 x + a_2 x^2 & (n=2) \\
\text{cubic:} \quad & y = a_0 + a_1 x + a_2 x^2 + a_3 x^3 & (n=3) \\
\text{general:} \quad & y = a_0 + a_1 x + a_2 x^2 + a_3 x^3 + \ldots + a_n x^n.
\end{aligned}
$$

The logarithmic function can be understood from the concept of a logarithm. From the identity $3^2 = 9$, we define the <u>exponent</u> 2 to be the logarithm of 9 to the <u>base</u> 3. This is written as $\log_3 9 = 2$. If $y = x^a$, then $a = \log_x y$. Note that the base x and the number y must always be positive. Two <u>bases</u> are commonly used, one is 10 and the other is the number e which is approximately equal to 2.718. The latter is called natural or Naperian logarithm and denoted by $\ln$, where the base e is not explicitly written out. The logarithmic operations follow several rules of which the following are often used

(i) log of a product: $\ln(xz) = \ln x + \ln z$ $x, z > 0$
(ii) log of a power: $\ln(x^a) = a \ln x$ $x > 0$

Example: $\ln(x^{-2}) = \ln(1/x^2) = \ln 1 - \ln(x^2)$
$= 0 - 2 \ln x$ $x > 0$
Since $\ln 1 = 0$ from the relation $e^0 = 1$

(iii) log of a quotient: $\ln(x/z) = \ln x - \ln z$ $x, z > 0$

Exponential functions, which arise nautrally as the solutions of differential equations describe the growth or decay. Thus if y is population and x is time in years and the annual rate of growth of population is 2 percent per year, then

$$y = y_0 e^{0.02x}$$

where y_0 is the initial size of the population. If instead the population is declining at the annual rate of 1 percent per year, then

$$y = y_0 e^{-0.01x}.$$

As economic examples, linear demand and supply functions describe a competitive market in equilibrium for a single good with its quantity x and price p

demand: $x_d = a_1 - b_1 p$ $a_1, b_1 > 0$
supply: $x_s = a_2 + b_2 p$ $a_2, b_2 > 0$
equilibrium: $x_d = x_s = x$.

On using the equilibrium condition, also called the market clearing condition one may easily obtain

$$(b_1 + b_2)p = a_1 - a_2.$$

If $(a_1 - a_2)$ is positive, one can solve for p and denote it as $\bar{p}$, the equilibrium price

$$\bar{p} = (a_1 - a_2)/(b_1 + b_2) > 0.$$

Substituting this value in the supply equation one obtains the equilibrium quantity $\bar{x}$ demanded and supplied

$$\bar{x} = (a_1 b_2 + a_2 b_1)/(b_1 + b_2) > 0.$$

Logarithmic functions are most commonly used in production function studies. A Cobb-Douglas production function views output (Y) as a function of two inputs, labor (L) and capital (K) in the following form

$$Y = H\, L^a\, K^b, \quad H = \text{a positive constant.}$$

Taking natural logarithms on both sides, one gets

$$\ln Y = \ln H + a\, \ln L + b\, \ln K.$$

Denoting by small letters the terms in natural logs, this can be written as

$$y = h + a\ell + bk.$$

Thus it is linear in logs. The exponents a and b are usually called the <u>elasticity</u> of output with respect to labor and capital respectively. On using the following conditions we define increasing, constant and decreasing returns to scale

$a + b > 1$ (increasing returns to scale)
$a + b = 1$ (constant returns to scale)
$a + b < 1$ (diminishing returns to scale).

As an example of exponential function, one may consider the utility function of a consumer $u(x)$ as

$$u(x) = 1 - e^{-ax}.$$

If the parameter a is positive, then it has positive marginal utility, $MU = ae^{-ax}$ and the marginal utility diminishes as the income or wealth variable x increases on the positive side. This type of function, with diminishing marginal utility is called a <u>concave</u> function, which has two most useful properties in economic theory. One is that this type of utility function has a maximum, that is usually finite. The second is that it exhibits risk aversion. Denoting marginal utility by $u'(x)$ and its rate of change by $u''(x)$, the absolute risk aversion measure R_A due to Arrow and Pratt is defined by $R_A = -u''(x)/u'(x)$ and in case of the exponential utility function above R_A takes the value $a > 0$. For a risk-taker (or gambler), R_A is negative, whereas a zero-value of R_A is on the borderline between risk-averse and risk-taking attitude. Thus the exponential utility function leads to a constant rate of absolute risk aversion, if the parameter a is a positive constant.

Most of the functions used in economics are <u>continuous</u>. Loosely speaking, if $y = f(x)$ is a continuous function, then the graph of y traces a smooth curve, when x varies in its domain e.g., a linear or quadratic function. Discontinuity occurs if the curve has breaks or sharp corners. More precisely, $y=f(x)$ is a continuous function of x within a certain domain of values of x, if y tends to a limiting value y_0, when x tends to apprach the value x_0 such that $y_0 = f(x_0)$. Thus $y = x/(1+x)$ is a continuous function of x for all finite real values of x, but $y = (2x-4)/(x^2-4)$ is not continuous at the value $x = 2$, since y is not defined at that value.

Another type of function is defined by the class of homogeneous functions. A production function with one output y depending on two inputs x_1, x_2 say

$$y = f(x_1, x_2)$$

is said to be homogeneous of degree $r \geq 0$, if multiplying each input by a positive number k results in an output level $k^r y$, that is

$$k^r y = f(kx_1, kx_2) \qquad k > 0.$$

For $r = 1$, the production function is homogeneous of degree one i.e., it is equivalent to constant returns to scale; hence doubling (trebling) each input makes output double (treble).

1.2.3 Vector and Matrix Algebra: Elementary Aspects

A vector is a list of scalars arranged in a certain <u>order</u>. The order matters; (4,3) is a different vector from (3,4). A vector is a row (column) vector, if it contains only one row (column) of scalars; (2,-1,3) is a row vector, but $\begin{pmatrix} 2 \\ -1 \\ 3 \end{pmatrix}$ is a column vector. A vector x with n elements $x_1, x_2, \ldots, x_n$ can thus be written in two ways, as a row vector or a column vector. Another way to view the vector x with n elements, a typical element being denoted by x_i ($1 \leq i \leq n$) is to consider it as a <u>point</u> in an n-dimensional rectangular coordinate system. For $n = 2$, this defines a point in a two-dimensional coordinate system, known as the Euclidean space.

A <u>matrix</u> is a list of vectors arranged in a certain order, where order matters; alternatively it is a two-dimensional array of scalars. The dimensionality of a matrix is measured by the number of rows and the number of columns it has. For example

$$\begin{bmatrix} 1 & 2 & 3 \\ 0 & -1 & 5 \end{bmatrix}$$

is a matrix with dimension 2x3, since it has two rows and three columns. Matrices are usually denoted by capital letters, with a typical element written with two subscripts, one for rows and the other for columns. For example, a matrix A with a typical scalar element (a_{ij}) has m rows and n columns (if $i=1,2,\ldots,m$ and $j=1,2,\ldots,n$). For $m=2$, $n=4$

$$A = \begin{bmatrix} a_{11} & a_{12} & a_{13} & a_{14} \\ a_{21} & a_{22} & a_{23} & a_{24} \end{bmatrix}.$$

Note that any n-element row (column) vector may be viewed as a matrix of dimension 1xn (nx1).

Like scalars, the vectors and matrices have their own rules of operation e.g., addition, multiplication, which are not always identical with those for scalars. From now on, we use vectors as column vectors only. If x is a column vector with dimension nx1, then its transpose denoted by x^T will denote a row vector of dimension 1xn. Likewise the matrix A above of dimension 2x3 will give a transpose, A^T of dimension 3x2.

For any two column vectors, $x = (x_i)$, $y = (y_i)$ each with n elements, the following rules of operation apply:

(i) Addition, $x + y = z$, where $z_i = x_i + y_i$; i=1,2,...,n.

If $x = \begin{pmatrix} 2 \\ 3 \end{pmatrix}$, $y = \begin{pmatrix} 0 \\ -1 \end{pmatrix}$, then $z = \begin{pmatrix} x_1 + y_1 \\ x_2 + y_2 \end{pmatrix} = \begin{pmatrix} 2 \\ 2 \end{pmatrix} = \begin{pmatrix} z_1 \\ z_2 \end{pmatrix}$.

Likewise for subtraction, $x - y = z$.

(ii) Product with a scalar, $\lambda x = \begin{bmatrix} \lambda x_1 \\ \lambda x_2 \\ \vdots \\ \lambda x_n \end{bmatrix}$

where λ is a scalar and x an n-element vector.

If $x = \begin{pmatrix} 2 \\ 3 \end{pmatrix}$, $\lambda = 2$, then $\lambda x = \begin{pmatrix} 4 \\ 6 \end{pmatrix}$.

(iii) Scalar or inner product of a row vector x^T and a column vector y, each with n elements,

$$x^T \cdot y = (x_1, x_2, \ldots, x_n) \begin{pmatrix} y_1 \\ y_2 \\ \vdots \\ y_n \end{pmatrix} = x_1 y_1 + x_2 y_2 + \ldots + x_n y_n$$

$$= \sum_{i=1}^{n} x_i y_i.$$

If $x^T = (2,3)$, $y = \begin{pmatrix} 1 \\ 2 \end{pmatrix}$, then $x^T \cdot y = x_1 y_1 + x_2 y_2 = (2 \times 1) + (3 \times 2) = 8$.

Note that there are more than one way of defining a product of two vectors and a scalar product is so named because the product results in a scalar. Further, if two vectors have unequal number of elements, then none of the operations of addition, subtraction or scalar product is defined. Thus the vectors must be conformable (i.e., <u>commutative</u>) in addition or scalar product. This is also true for the matrices.

For any two matrices $A = (a_{ij})$, $B = (b_{ij})$ of dimension or order mxn, the following operations are valid

(i) Sum: $A + B = C$, where $c_{ij} = a_{ij} + b_{ij}$; $i=1,2,\ldots,m$; $j=1,2,\ldots,n$.

If $A = \begin{bmatrix} 1 & 2 \\ 4 & 3 \end{bmatrix}$, $B = \begin{bmatrix} 4 & 1 \\ 3 & 2 \end{bmatrix}$, then $C = \begin{bmatrix} c_{11} & c_{12} \\ c_{21} & c_{22} \end{bmatrix} = \begin{bmatrix} 5 & 3 \\ 7 & 5 \end{bmatrix}$.

(ii) Scalar product: $\lambda A = (\lambda a_{ij})$, for all i and j.

If $A = \begin{bmatrix} 1 & 2 \\ 4 & 3 \end{bmatrix}$, $\lambda = 3$ then $\lambda A = \begin{bmatrix} 3 & 6 \\ 12 & 9 \end{bmatrix}$.

(iii) Product of A (mxn) and B (nxm) yields a matrix C = AB of order mxm. Two conditions must hold before a product of two matrices results. One is that the number of <u>columns</u> of the first matrix (A) must equal the number of <u>rows</u> of the second matrix (B), before the two matrices A and B in this order AB are conformable in multiplication. Second, the rule of multiplication is that the row i of the first matrix (A) must be multiplied by column j of second matrix (B), element by element and then added to form the (ij) element of the product matrix C, where the element c_{ij} of C is given by

$$c_{ij} = \sum_{k=1}^{n} a_{ik} b_{kj}, \text{ for all } i,j = 1,2,\ldots,n.$$

If $A = \begin{bmatrix} 1 & 2 \\ 4 & 3 \end{bmatrix}$, $B = \begin{bmatrix} 1 & 2 & 3 \\ 0 & 1 & 1 \end{bmatrix}$

then C = AB is of order 2x3 and

$c_{11} = (1 \times 1) + (2 \times 0) = 1$; $c_{12} = (1 \times 2) + (2 \times 1) = 4$; $c_{13} = (1 \times 3) + (2 \times 1) = 5$
$c_{21} = (4 \times 1) + (3 \times 0) = 4$; $c_{22} = (4 \times 2) + (3 \times 1) = 11$; $c_{23} = (4 \times 3) + (3 \times 1) = 15$

$C = \begin{bmatrix} 1 & 4 & 5 \\ 4 & 11 & 15 \end{bmatrix}$.

In this case the product BA is not defined, since in this order the two matrices are not conformable, i.e., the number of columns of B is not equal to the number of rows of A.

(iv) Some special matrices

(a) A <u>diagonal</u> matrix is a square matrix with zeros everywhere, except on the main diagonal e.g.,

$$\begin{bmatrix} a_{11} & 0 & 0 \\ 0 & a_{22} & 0 \\ 0 & 0 & a_{33} \end{bmatrix}.$$

(b) An <u>identity</u> matrix I_n of order n is a diagonal matrix of order nxn with 1 in the main diagonal and zeros elsewhere. This matrix I_n or, sometimes denoted by I plays the same role in matrix algebra as the number one in real scalar number system e.g.,

$$Ix = x: \quad \begin{bmatrix} 1 & 0 \\ 0 & 1 \end{bmatrix} \begin{pmatrix} x_1 \\ x_2 \end{pmatrix} = \begin{pmatrix} x_1 \\ x_2 \end{pmatrix}.$$

$I_n A = A$ where $A = (a_{ij})$ is of order nxn.

(c) The inverse of a square matrix A denoted by A^{-1} is obtained from the identity $A^{-1}A = I_n = AA^{-1}$. For the inverse matrix to be defined, its determinant denoted by |A| must not be zero. A square matrix whose determinant is zero is called <u>singular</u>. The determinant of

$$A = \begin{bmatrix} 1 & 2 \\ 4 & 3 \end{bmatrix} \text{ is } |A| = (3 \times 1) - (4 \times 2) = 3 - 8 = -5$$

and in general if

$$A = \begin{bmatrix} a_{11} & a_{12} & a_{13} \\ a_{21} & a_{22} & a_{23} \\ a_{32} & a_{32} & a_{33} \end{bmatrix}$$

then

$$A = (-1)^{1+1} a_{11} \begin{vmatrix} a_{22} & a_{23} \\ a_{32} & a_{33} \end{vmatrix} + (-1)^{1+2} a_{12} \begin{vmatrix} a_{21} & a_{23} \\ a_{31} & a_{33} \end{vmatrix}$$

$$+ (-1)^{1+3} a_{13} \begin{vmatrix} a_{21} & a_{22} \\ a_{31} & a_{32} \end{vmatrix}$$

$$= a_{11}(a_{22}a_{33} - a_{23}a_{32}) - a_{12}(a_{21}a_{33} - a_{31}a_{23})$$

$$+ a_{13}(a_{21}a_{32} - a_{31}a_{22}).$$

Here we have expanded $|A|$ in terms of the first row e.g., we take each element of the first row and then multiply it by its appropriate subdeterminant, taking care of the sign and then add; thus the second element a_{12} of the first row has the negative sign since $(-1)^{1+2} = -1$, where the two indices of a_{12} are added. Thus $|A|$ is the linear sum of products, where each product is obtained by multiplying an element a_{ij} by its appropriate subdeterminant and then determining the sign of the product from $(-1)^{i+j}$. For the element a_{ij}, the subdeterminant is obtained by deleting row i and column j from the original matrix A. These subdeterminants are called <u>minors</u> and signed minors are called <u>cofactors</u>. Thus the minor of the element a_{12} is the subdeterminant $\begin{vmatrix} a_{21} & a_{23} \\ a_{31} & a_{33} \end{vmatrix}$ but its cofactor is $-(a_{21}a_{33} - a_{31}a_{23})$.

Since each element of the square matrix A has a cofactor i.e., a_{ij} has cofactor denoted by $|C_{ij}|$ say, one could construct a co-factor matrix C. Thus for the 3×3 matrix A, its cofactor matrix is

$$C = \begin{bmatrix} |C_{11}| & |C_{12}| & |C_{13}| \\ |C_{21}| & |C_{22}| & |C_{23}| \\ |C_{31}| & |C_{32}| & |C_{33}| \end{bmatrix}$$

The transpose of this matrix C^T is termed the <u>adjoint</u> of A and is denoted by Adj A. An alternative definition of the inverse A^{-1} of a nonsingular square matrix is as follows

$$A^{-1} = \frac{Adj\ A}{|A|}, \qquad A \neq 0$$

There exist two important uses of the inverse matrix in economics. One is in solving a set of two or more linear equations and the second in series expansions. As an example of the first, consider the following demand-supply model where x_1, x_2 are the quantity and unit price of a good for which supply equals demand

demand: $x_1 = b_1 - a_{12} x_2$

supply: $x_1 = b_2 + a_{22} x_2$.

$a_{ij} > 0$, all $i,j=1,2$

Rearranging this can be written as

$$\begin{bmatrix} 1 & a_{12} \\ 1 & -a_{22} \end{bmatrix} \begin{pmatrix} x_1 \\ x_2 \end{pmatrix} = \begin{pmatrix} b_1 \\ b_2 \end{pmatrix}.$$

If the coefficient matrix is nonsingular i.e. $A = -a_{22} - a_{12} \neq 0$, then the solution of the equilibrium quantity and price can be obtained by the inverse

$$\begin{pmatrix} x_1 \\ x_2 \end{pmatrix} = \begin{bmatrix} 1 & a_{12} \\ 1 & -a_{22} \end{bmatrix}^{-1} \begin{pmatrix} b_1 \\ b_2 \end{pmatrix}$$

$$= \begin{bmatrix} -a_{22}/|A| & a_{12}/|A| \\ 1/|A| & 1/|A| \end{bmatrix} \begin{pmatrix} b_1 \\ b_2 \end{pmatrix}$$

where the determinant A of the coefficient matrix associated with the vector $x = \begin{pmatrix} x_1 \\ x_2 \end{pmatrix}$ equals $-(a_{22} + a_{12})$. The above equation system can also be solved by

<u>Cramer's</u> rule, when each element of the vector x has to be solved. By this rule

$$x_1 = \frac{\begin{vmatrix} b_1 & a_{12} \\ b_2 & -a_{22} \end{vmatrix}}{|A|} = \frac{-(a_{22}b_1 + a_{12}b_2)}{-(a_{22} + a_{12})}$$

$$x_2 = \frac{\begin{vmatrix} 1 & b_1 \\ 1 & b_2 \end{vmatrix}}{|A|} = \frac{b_2 - b_1}{-(a_{12} + a_{22})}.$$

In the two cases of x_1 and x_2, the denominator $|A|$ is the same but the numerator follows the rule: replace that column of the coefficient matrix A by the right-hand side vector $b = \begin{pmatrix} b_1 \\ b_2 \end{pmatrix}$ for which the subscript of x_i is required to be solved, i.e. for x_1, we replace the first column of the coefficient matrix.

As an example of the second type of use, consider an open-static Leontief-model of an economy with two branches (or sectors) having gross outputs x_1 and x_2. The output of each sector, in demand-supply equilibrium, is partly used as raw materials (inputs) by other sectors and partly for final demand by households. Thus

$$x_1 = a_{11}x_1 + a_{12}x_2 + d_1$$
$$x_2 = a_{21}x_1 + a_{22}x_2 + d_2.$$

In matrix terms

$$x = A x + d$$

i.e. $(I - A) x = d$,

$$I = \text{identity matrix} = \begin{bmatrix} 1 & 0 \\ 0 & 1 \end{bmatrix}.$$

The input-coefficient matrix $A = [a_{ij}]$ here has several properties known as <u>Leontief-properties</u> e.g. each a_{ij} is nonnegative, though all the elements cannot be identically zero and the sum of each column of the coefficient matrix has to be less than one i.e. $a_{11} + a_{21} < 1$, $a_{12} + a_{22} < 1$. The second condition means that each branch is productive in the sense that its gross output level leaves some positive surplus after raw material costs are deducted. Under these two conditions the matrix (I-A) has an inverse as follows

$$[I-A]^{-1} = [I + A + A^2 + \ldots + \emptyset]$$

where $\emptyset$ is the matrix with each element zero; if the demand vector d has positive elements then the gross output vector x would also be positive:

$$x = [I-A]^{-1} d = [I + A + A^2 + \ldots + \emptyset] d$$

The inverse matrix $B = [I-A]^{-1}$ is also called the multiplier matrix associated with demand d, and it gives the equilibrium quantity of supply of output. If expected output $\hat{x}$ next year is not equal to this equilibrium output x, we have <u>disequilibrium</u>. One of the consequences of disequilibrium may be felt in price increases or decreases in some sectors.

1.2.4 Random Numbers in Economics

A random number is most often applied in economics for three purposes. One is to test how good a particular hypothesis e.g. a demand function is against observed empirical data. Thus a deterministic demand function $y = a-bx$, y = quantity and x = price is written as a stochastic relation

$$y = a - bx + e$$

where e is a disturbance or error term with some probability structure. Methods of econometrics are then applied to estimate the coefficients or parameters a,b. For example, the least squares method determines the estimates $(\hat{a},\hat{b})$ by minimizing the sum $\sum_{t=1}^{N} e_t^2$ of squared errors over the observed samples $t=1,2,\ldots,N$ of (y_t,x_t). If the estimates $(\hat{a},\hat{b})$ satisfy some optimality criteria e.g. unbiasedness and minimum variance, then the relation

$$y = \hat{a} - \hat{b}x$$

is considered a good approximation of the theoretical hypothesis $y=a-bx$, provided the fit is good.

A second use of the concept of a random or stochastic element is in characterizing the riskiness of an outcome or decision. Let x be a (discrete) random variable taking values $x_1, x_2, \ldots, x_n$ each with probability $p_1, p_2, \ldots, p_n$ where $p_i \geq 0$ and $\sum_{i=1}^{n} p_i = 1$. Then the mean m and variance v of x is computed as $m = \sum_{i=1}^{n} p_i x_i$ and $v = \sum_{i=1}^{n} p_i (x_i - m)^2$. Riskiness is sometimes associated with high variance. Thus an investor when comparing two securities in terms of their returns may usually consider the first security investment to be more risky if it has higher variance of returns with identical mean returns for both. Risk aversion is a term used to specify the class of investors who are more cautious in the sense that they choose less risky securities or portfolios.

Another use of random numbers is the area of simulation using a computer, either a regular or a personal computer. Artificial data with given mean and variances may be generated e.g. a normal distribution may be used to test the empirical fit of a linear demand function. Hidden nonlinearities in specification or statistical estimation may thus be tested. Also the degree of robustness of a particular estimate i.e. its insensitivity to data variations may be evaluated. Recent applications of large scale econometric and national economic models stress such robustness.

1.3 Some Advanced Concepts

The following advanced concepts are from the theory of functions and matrices.

(i) Continuous functions and their uses in Taylor series expansion, maximizing and minimizing and in implicit function differentiation.

(ii) Matrices and their applications in eigenvalues, quadratic forms and some basic theorems useful in applied economics.

1.3.1 Continuous Functions: Their Uses

The idea of a continuous function $y=f(x)$ is based on the notion of a limit. Thus we ask if the point x gets closer to x_0, does $f(x)$ get closer to some number L? If it does, then the function has a limit L. The function $f(x)$ is defined to be continuous at the point x_0 if three conditions simultaneously hold: (i) $f(x_0)$ exists, (ii) $\lim_{x \to x_0} f(x)$ exists and (iii) $\lim_{x \to x_0} f(x) = f(x_0)$. For instance, the quadratic function $f(x) = 1 + x + 2x^2$ is continuous at $x_0 = 0$, since $f(x_0) = f(0) = 1$ exists, the limit $f(x)$ exists and $\lim_{x \to 0} f(x) = f(0) = 1$. But suppose we take a function which has jumps at $x = x_0 = 0$:

$$f(x) = \begin{cases} 1 + 2x^2 & \text{for } x \neq 0 \\ 2 & \text{for } x = 0 \end{cases}$$

then for $x \neq 0$ $\lim f(x) = 1$ from the first part but $f(0) = 2$ and $2 \neq 1$; hence this function is not continuous at $x = 0$.

Derivatives

Most continuous functions possess derivatives, although there are exceptions when there are kinks in the graph as in linear programming. The derivative of a function $f(x)$ at a point x_0 is given by

$$\frac{df(x_0)}{dx} = \lim_{t \to 0} \frac{f(x_0 + t) - f(x_0)}{t}$$

if this limit exists, in which case $f(x)$ is differentiable at x_0. Often we use the notation $df(x_0)/dx = f'(x_0)$ and also the differential operator D for d/dx where $D(f(x_0)) = f'(x_0)$. Note that for the limit to exist at the point x_0, it is necessary that the function be continuous at that point.

Rules for differentiation of various functional forms are given in any text on calculus. The following are most often used in economics

(1) If $f(x) = a$, a is a constant, then $D(f(x)) = f'(x) = 0$.

(2) If $f(x) = ax^n$, a and n are nonzero constants then $f'(x) = anx^{n-1}$ e.g. $f(x) = x^3$, then $f'(x) = 3x^2$.

(3) Sum of two functions
If $f(x) = g(x) + h(x)$, then $f'(x) = g'(x) + h'(x)$.

(4) Product of two functions
If $f(x) = h(x) \cdot g(x)$ then $f'(x) = h(x) \cdot g'(x) + h'(x) \cdot g(x)$ e.g. $h(x) = x^2$, $g(x) = x$, $f(x) = x^2 \cdot x = x^3$ then $f'(x) = 3x^2$ and also $f'(x) = x^2 \cdot 1 + 2x \cdot x = 3x^2$.

(5) Ratio of two functions

If $f(x) = \frac{g(x)}{h(x)}$, then $f'(x) = \frac{g'(x)h(x) - g(x)h'(x)}{(h(x))^2}$

e.g. $f(x) = \frac{x^2}{x^6}$, $f'(x) = \frac{2x(x^6) - x^2(6x^5)}{(x^6)^2} = \frac{-4x^7}{x^{12}} = -4x^{-5}$.

(6) Logarithmic function
If $f(x) = \log x$, then $f'(x) = 1/x$.

(7) Exponential function
If $f(x) = e^x$, then $f'(x) = e^x$
e.g. $f(x) = e^{3x}$, $f'(x) = 3e^{3x}$.

(8) Chain rule
This rule is helpful in finding the rate of change in y w.r.t. changes in x, when y depends on another variable z and z depends on x.
If $y = g(z)$ and $z = h(x)$, then $y = g\{h(x)\}$ i,e., y is a function of a function. Here

$$f'(x) = \frac{dg\{h(x)\}}{dh} \cdot \frac{dh(x)}{dx} .$$

e.g., let $f(x) = e^{h(x)}$, $h(x) = 3 + 2x$
$g\{h(x)\} = e^{h(x)} = e^{3+2x}$. Then

$\frac{dg\{h(x)\}}{dh} = \frac{de^h}{dh} = e^h$ and $\frac{dh(x)}{dx} = \frac{d(3 + 2x)}{dx} = 2$

so $f'(x) = e^h \cdot 2 = 2e^{3+2x}$.

(9) Higher order derivatives

If $f(x) = x^4$, then $D(f) = f'(x) = 4x^3$ and continuing $f''(x) = D(f') = D^2(f) = 12x^2$, $f'''(x) = D^3(f) = 24x$, $D^4(f) = 24$ and $D^5(f) = 0$.

1.3.2 Taylor Series Expansion

Taylor series expansion of a smooth function $f(x)$ can be thought of as a way of approximating the function at some point x by using a known point x_0 and its derivatives at that point.

Taylor's Theorem

Let $f(x)$ be a continuous function with continuous derivatives upto the $(n+1)$-th order on the closed interval from x_0 to x, then $f(x)$ can be expanded as follows

$$f(x) = f(x_0) + \frac{df(x_0)}{dx}(x - x_0) + \frac{1}{2!}\frac{d^2f(x_0)}{dx^2}(x - x_0)^2 + \ldots + \frac{1}{n!}\frac{d^nf(x_0)}{dx^n}(x - x_0)^n + R_n$$

where R_n the remainder is given by

$$R_n = \frac{1}{(n+1)!}\frac{d^{(n+1)}f(y)}{dx^{(n+1)}}(x - x_0)^{n+1}$$

for some point y in the closed interval $x_0 \leq y \leq x$.

It is clear that if $R_n \to 0$ as $n \to \infty$, then this theorem allows $f(x)$ to be expressed as an infinite series of terms. If x_0 is taken as zero, then we obtain

$$f(x) = f(0) + f'(0)x + \frac{1}{2!}f''(0)x^2 + \ldots + R_n$$ where $\frac{d^nf(x)}{dx^n}$ is the n-th order

derivative of $f(x)$ and $n! = 1,2,3,\ldots,n$ is the factorial n.

Applications

(a) Let $f(x) = (1+x)^3$. We know by direct multiplication that $(1+x)^3 = 1 + 3x + 3x^2 + x^3$. To confirm this we expand $f(x)$ in Taylor series around $x_0 = 0$, noting that the derivatives $\frac{d^nf(x_0)}{dx}$ of order 4 (i.e., n=4) and above is zero i.e., then

$$f(x) = 1 + 3x + \frac{1}{2!}6x + \frac{1}{3!}6x^2 = 1 + 3x + 3x^2 + x^2$$, since $f'(x) = 3(1+x)^2 = 3$ at $x=0$, $f''(x) = 6(1+x) = 6$ at $x=0$, $f'''(x) = 6$ at $x=0$ and $f^{(iv)}(x) = 0$ at $x=0$.

(b) A nonlinear differentiable function $f(x)$ has a local maximum or minimum at a point x_0 if it satisfies the necessary or first-order condition

$$\frac{df(x_0)}{dx} = 0.$$

A sufficient condition for x_0 to be a local maximum (minimum) is that $f(x)$ is a concave (convex) function in a neighborhood $N(x_0)$ of x_0. If $f(x)$ is concave around x_0, then $-f(x)$ is convex around x_0; hence we may only discuss the case of concavity. There are three simple ways to characterize the concavity of a function. One is the second derivative rule. If $f''(x_0) = \frac{d^2 f(x_0)}{dx^2}$ is negative, then $f(x)$ is strictly concave (if $f''(x_0) \leq 0$, it is weakly concave or simply concave). The second is the tangent rule, which applies when the second derivatives do not exist. For a concave once differentiable function, the tangent line lies on or above the curve $y = f(x)$ i.e.

$$f(x_0) + f'(x_0)(x - x_0) \geq f(x). \tag{1}$$

The left-hand side is the equation of the tangent line with slope $f'(x_0)x$ and intercept $(f(x_0) - f'(x_0)x_0)$, obtained from the Taylor series expansion of $f(x)$ upto linear terms. The sign of strict inequality ($>$) is required for strict concavity, otherwise it is weakly concave or simply concave. For convex functions the sign $\geq$ is reversed as $\leq$. The neighborhood or domain $N(x_0)$ around the point x_0 may be small or large. In the latter case we talk of global maximum or minimum. A third way to define a concave function is by the chord rule. Let $x_1 \leq x \leq x_2$ be an interval on the x-axis where x_1, x_2 are two distinct endpoints. Define a weighted average point $\bar{x} = wx_1 + (1-w)x_2$ for any w satisfying $0 \leq w \leq 1$ and consider the ordinate $f(\bar{x})$. If $f(\bar{x})$ satisfies

$$f(\bar{x}) \geq w f(x_1) + (1-w) f(x_2)$$

then the function is concave (strictly concave if the inequality sign is strict i.e. $>$) in the interval $I = \{x \mid x_1 \leq x \leq x_2\}$. In verbal terms the chord given by the line $w f(x_1) + (1-w) f(x_2)$ is on or below the ordinate $f(\bar{x})$ at $\bar{x}$. Note that for the chord rule we do not need the existence of even the first derivative of the function.

(c) For a competitive firm selling a good x with a given price p, profit z may be written as $z = px - c(x)$, where px is total revenue and $c(x)$ is total cost. If $c(x)$ is a strictly convex function in x, then $z = z(x)$ is strictly concave. Hence the first order stationarity condition

$$\frac{dz(x)}{dx} = 0 = p - \frac{dc(x)}{dx}$$

yields an x_0 at which price p equals marginal cost $dc(x)/dx$ and this yields a sufficient condition for the maximum of $z(x)$ i.e. $z(x_0) > z(x)$. Since this holds for all x, as there is no restriction on the domain of x we have a global maximum.

(d) Taylor series expansion also holds for a smooth function of many variables, except that the partial derivatives have to be used in place of the total derivatives. Given a function $y = f(x_1, x_2, \ldots, x_n)$ of n variables, which may also be denoted by $y = f(x)$ where x is an n-element column vector and y is a scalar, we define the partial derivative of $f(x)$ with respect to x_i at the point x^o by

$$\frac{\partial f(x^o)}{\partial x_i} = \lim_{t \to 0} \frac{f(x_1^o, \ldots, x_i^o + t, \ldots, x_n^o) - f(x_1^o, \ldots, x_i^o, \ldots, x_n^o)}{t}$$

if this limit exists. It is also denoted by $f_{x_i}(x^o)$ or, $f_i(x^o)$ which really means taking the derivative of $f(x^o)$ with respect to x_i, keeping all other x's constant. Thus for a quadratic function in two variables $f(x_1, x_2) = 2 + x_1 + x_2 - x_1^2 - 2x_2^2 + x_1 x_2$ we get

$$f_1(x) = \frac{\partial f(x)}{\partial x_1} = 1 - 2x_1 + x_2; \quad f_2(x) = 1 - 4x_2 + x_1$$

Differentiating once again

$$f_{11}(x) = \frac{\partial}{\partial x_1}\left(\frac{\partial f(x)}{\partial x_1}\right) = -2, \quad f_{12}(x) = \frac{\partial}{\partial x_1}\left(\frac{\partial f(x)}{\partial x_2}\right) = 1$$

$$f_{21}(x) = \frac{\partial}{\partial x_2}\left(\frac{\partial f(x)}{\partial x_1}\right) = 1, \quad f_{22}(x) = \frac{\partial}{\partial x_2}\left(\frac{\partial f(x)}{\partial x_2}\right) = -4$$

The matrix H defined by the second partial derivatives

$$H(x) = \begin{bmatrix} f_{11}(x) & f_{12}(x) & \cdots & f_{1n}(x) \\ \vdots & \vdots & & \vdots \\ f_{n1}(x) & f_{n2}(x) & \cdots & f_{nn}(x) \end{bmatrix}$$

is called the Hessian matrix which now takes the place of second derivative of a function $f(x)$ of one variable x. In the above quadratic example

$$H = \begin{bmatrix} -2 & 1 \\ 1 & -4 \end{bmatrix}$$

where note $f_{12}(x) = f_{21}(x)$. The equality of cross partials holds generally i.e. $f_{ij}(x) = f_{ji}(x)$ and the Hessian matrix is symmetric.

The Taylor series expansion of f(x) around the vector point x^o may now be written as

$$f(x) = f(x^o) + \sum_{i=1}^{n} \frac{\partial f(x^o)}{\partial x_i} (x_i - x_i^o)$$

$$+ \frac{1}{2!} \sum_{i=1}^{n} \sum_{j=1}^{n} \frac{\partial^2 f(x^o)}{\partial x_i \partial x_j} (x_i - x_i^o)(x_j - x_j^o)$$

+ higher order terms.

This can also be written in vector notation as

$$f(x) = f(x^o) + f_x(x^o)^T (x - x^o) + \frac{1}{2!} (x - x^o)^T H(x^o)(x - x^o) + \text{higher order terms}$$

where T denotes transpose, $f_x(x^o)$ is the gradient vector $\partial f/\partial x$ and $H(x^o)$ is the Hessian matrix evaluated at x^o.

1.3.3 Hessian Matrix and Concavity

As in the scalar case, the necessary conditions for the maximum or minimum of a scalar differentiable function f(x) of vector x are given by

$$\frac{\partial f(x^o)}{\partial x_i} = 0, \quad i=1,2,\ldots,n.$$

A sufficient condition for the vector x^o to be a maximum is stated by any one of the three conditions.

(i) The Hessian matrix $H(x^o) = (\frac{\partial^2 f(x^o)}{\partial x_i \partial x_j})$ of second partials evaluated at x^o is negative definite i.e. for any n-element nonzero vector u, the scalar quantity $u^T H(x^o) u$ is negative (for a nonunique maximum, $H(x^o)$ has to be negative semidefinite in the sense $u^T H(x^o) u \leq 0$). An alternative definition is given below.

(ii) The function f(x) is concave around x^o in the sense

$$f(x^o) + f_x(x^o)^T \cdot (x - x^o) \geq f(x)$$

where $f_x(x^o)$ is called the gradient vector with elements $f_{x_i}(x^o) = f_i(x^o) = \frac{\partial f(x^o)}{\partial x_i}$, $i=1,\ldots,n$. Again for strict concavity the inequality sign above has to be strict (i.e. >).

(iii) The function satisfies the chord rule:

$$f(\bar{x}) \geq w\, f(x^1) + (1-w)\, f(x^2),\ 0 \leq w \leq 1$$

where x^1, x^2 are two vector points such that $\bar{x} = w\, x^1 + (1-w)\, x^2$ and the inequality sign holds in the domain $I = \{x \mid x^1 \leq x \leq x^2\}$.

Consider some examples:

(a) The quadratic utility function

$$u(x) = x_1 + 2x_2 - 0.5\,(x_1^2 + x_2^2)$$

is strictly concave, since the Hessian matrix is

$$H = \begin{bmatrix} -1 & 0 \\ 0 & -1 \end{bmatrix}$$

and choose u as the column vector $\begin{pmatrix}1\\1\end{pmatrix}$ to obtain $u^T H u = -2 < 0$.

(b) The function $f(x) = 1 - e^{-a_1 x_1 - a_2 x_2}$ for nonzero, a_1 and a_2 is concave but not strictly concave, since the Hessian matrix is

$$H = \begin{bmatrix} -a_1^2 & -a_1 a_2 \\ -a_1 a_2 & -a_2^2 \end{bmatrix} (e^{-a_1 x_1 - a_2 x_2})$$

and it is singular because it's determinant is zero.

An alternative definition of negative definiteness of a symmetric matrix H is as follows. Staying along the main diagonal, we form submatrices of orders $1, 2, \ldots$, upto n. Let $D_1, D_2, \ldots, D_n$ be their determinantal values. If $D_1 < 0$ and thereafter $D_2, D_3, \ldots, D_n$ alternate in sign i.e., $D_2 > 0$, $D_3 < 0$ and so on, then the matrix H is negative definite. It is negative semidefinite if $D_1 \leq 0$ and thereafter $D_2 \geq 0$, $D_3 \leq 0$ and so on. It is easy to check in Ex. 2 that $D_1 = -a_1^2 < 0$, $D_2 = 0$ and hence H is negative semidefinite i.e. the underlying function is concave but not strictly concave.

1.3.4 Implicit Functions

So far we have considered functions which are explicit in the sense that the dependent variable $y = f(x)$ can be <u>explicitly</u> written as a function of x. But in many economic problems the function may be implicit e.g.

$$f(x,y) = y + 2x^2 y - 5 = 0 \tag{1.1}$$

Here for every value of x one can find a value for y satisfying this equation. But if we have an equation as

$$f(x,y) = (x-1) - (y-2)^2 = 0 \tag{1.2}$$

we get for each value of x, two values of y e.g., for $x = 2$, $y = 3$ and 1. The implicit function theorem states the general conditions under which an equation of the form

$$f(x_1 x_2, y) = 0 \tag{2}$$

may be explicitly expressed as

$$y = g(x_1, x_2) \tag{3}$$

for some suitable function $g(\cdot)$. This theorem applies to n independent variables $x_1, \ldots, x_n$, although we state it for two.

Suppose the function f in (2) has continuous partial derivatives f_1, f_2, f_y and $f_y = \partial f/\partial y$ is not zero at a vector point $x^o = (x_1^o, x_2^o)$ satisfying (2) then there exists a two dimensional neighborhood $N(x^o)$ of the point where y can be explicitly written in the form (3). Since $\partial f(x,y)/\partial y = 1 + 2x^2$ is not zero in (1.1) for any $x = x_o$, the explicit function for (1.1) becomes

$$y = 5/(1 + 2x^2)$$

For the function $f(x,y)$ in (1.2), $f_y = -2(y - 2) = 0$ for $y = 2$ but $f_y \neq 0$ for $y \neq 2$. Hence for $y \neq 2$ the implicit function theorem applies and we get two separate functions

$$y = 2 + \sqrt{x - 1} \text{ and } y = 2 - \sqrt{x - 1}$$

hence two separate derivatives dy/dx can be found respectively as

$$\frac{dy}{dx} = (1/2)(x - 1)^{-1/2} \; ; \; (y \neq 2)$$

and

$$\frac{dy}{dx} = -(1/2)(x - 1)^{-1/2} \; ; \; (y \neq 2)$$

We may note two important applications of the implicit function theorem. One is that it allows us to evaluate the partial derivatives $\partial y/\partial x_i$ $i=1,2,\ldots,n$ so that the impact of varying any one x_i keeping all other x_j's unchanged may be evaluated. This is most useful in comparative static analysis in economic models. Secondly, one can apply Taylor series expansion to linearize a nonlinear function such as (2) if there exists a neighborhood around a point $x^o = (x_1^o, x_2^o)$ where the implicit function rule holds. This is useful in linearizing nonlinear equations and obtaining a linear approximate solution and hence analyzing its sensitivity.

To illustrate the first application, consider (2) as an implicit production function with one output y and two inputs x_1 = labor and x_2 = capital say. If the three partial derivatives f_1, f_2, f_y exist one can express (2) in terms of the differentials dx_1, dx_2, dy as

$$f_1 dx_1 + f_2 dx_2 + f_y dy = 0. \qquad (4)$$

Assume that only x_2 varies but not x_1 so that dx_1 is zero, then we get $0 + f_2 dx_2 + f_y dy = 0$ i.e.

$$\left. \frac{dy}{dx_2} \right|_{x_1 = \text{constant}} = \partial y / \partial x_2 = \frac{-f_2}{f_y}.$$

Likewise if x_1 varies but x_2 = constant so that $dx_2 = 0$ but not dx_1, then

$$f_1 dx_1 + f_y dy = 0 \text{ i.e. } \left. \frac{dy}{dx_1} \right|_{x_2 = \text{constant}} = \partial y / \partial x_1 = \frac{-f_1}{f_y}.$$

In the general case $f(x_1, x_2, \ldots, x_n, y) = 0$ if the implicit function theorem holds we would have

$$\frac{\partial y}{\partial x_i} = -f_i / f_y; \quad i = 1, 2, \ldots, n.$$

Note also that from (4) we can derive the equation of an <u>isoquant</u>, where output y is held constant i.e. dy=0 by the compensating variations in x_1 and x_2 i.e.

$$f_1 dx_1 + f_2 dx_2 + 0 = 0.$$

This yields

$$\left. \frac{dx_2}{dx} \right|_{y = \text{constant}} = \frac{\partial x_2}{\partial x_1} = -\frac{f_1}{f_2}$$

which measures the negative slope of an isoquant since the marginal products f_1, f_2 are usually positive.

For the second application, assume that we have a demand supply model of a competitive market for one good, where demand is $d = f(p)$ but the supply quantity s is given by a function $s = g(p-t)$ of net price i.e. price p less a tax of t cents per unit sold. The functions f and g are continuous and nonlinear and we assume that an equilibrium price exists at which the market clears i.e.

$$f(p) = g(p-t) \tag{5}$$

the consumers want to buy exactly what the suppliers sell. This equation implicitly defines the equilibrium p, say $\bar{p}$ as a function of the tax rate t. If the conditions of the implicit function theorem hold, then we can explicitly obtain $\bar{p} = h(t)$ as a function of t and hence the marginal impact $\partial\bar{p}/\partial t = d\bar{p}/dt$ of a change in t. Thus let us assume the implicit function theorem holds and define $\hat{p}(t) = p(t) - t$, where p is written as $p(t)$ to indicate its implicit dependence on t. Differentiating by the chain rule we obtain from $f(p(t)) = g\{\hat{p}(t)\}$

$$\frac{df(p)}{dp} \cdot \frac{dp}{dt} = \frac{dg(\hat{p})}{d\hat{p}} \cdot \frac{d\hat{p}}{dt} = \frac{dg(\hat{p})}{d\hat{p}} \cdot (\frac{dp}{dt} - 1).$$

Hence

$$\frac{dp}{dt} = \frac{-g_{\hat{p}}}{f_p - g_{\hat{p}}}, \text{ where } \quad g_{\hat{p}} = dg/d\hat{p} \\ f_p = df/dp.$$

Clearly for demand functions $f_p < 0$ since the demand curve has a negative slope and for supply functions $g_{\hat{p}} > 0$, since the supply curve has a positive slope.

Hence $dp/dt = g_{\hat{p}}/(|f_p| + g_{\hat{p}})$, where $|f_p|$ = absolute value of f_p implying $0 < dp/dt < 1$ so that the equilibrium price goes up by an increase in tax rate but by an amount less than the tax.

1.4 Advanced Concepts in Matrix Algebra

Matrices provide a convenient shorthand notation for a system of equations; they also generalize the notion of a vector, since a matrix can be viewed as a collection of vectors. But there are other basic operations which are clarified by matrix algebra in a more fundamental way than mere notational abbreviations. Three of these aspects would now be illustrated: (i) the concept of rank of a matrix and linear dependence of a set of vectors, (ii) the notions of eigenvalues (also called characteristic values) and eigenvectors (also called characteristic vectors) and their applications in quadratic forms and difference or differential equations, and (iii) special matrices like the Frobenius and Metzler matrix that are often used in economic models.

1.4.1 Rank and Linear Independence of Vectors

Consider a 3 by 3 matrix A

$$A = \begin{bmatrix} 1 & 0 & 2 \\ 3 & 1 & 2 \\ 0 & 1 & 1 \end{bmatrix}. \tag{1}$$

It can be written as a collection of 3 column vectors $(a_1 \ a_2 \ a_3)$, where
$a_1 = \begin{pmatrix} 1 \\ 3 \\ 0 \end{pmatrix}$, $a_2 = \begin{pmatrix} 0 \\ 1 \\ 1 \end{pmatrix}$, $a_3 = \begin{pmatrix} 2 \\ 2 \\ 1 \end{pmatrix}$ or, a collection of 3 row vectors $\begin{pmatrix} b_1 \\ b_2 \\ b_3 \end{pmatrix}$ where
$b_1 = (1 \ 0 \ 2)$, $b_2 = (3 \ 1 \ 2)$, $b_3 = (0 \ 1 \ 1)$. The rank of this square matrix can be defined in two simple ways. One is by the highest order determinant which is nonzero. The highest order here is three and the determinantal value $A = 5$ which is not zero. Hence the rank of A denoted by $r(A) = 3$. If A were

$$A = \begin{bmatrix} 1 & 0 & 0 \\ 0 & 1 & 0 \\ 0 & 0 & 0 \end{bmatrix}, \quad \text{then } r(A) = 2$$

since there exists a subdeterminant $\begin{pmatrix} 1 & 0 \\ 0 & 1 \end{pmatrix}$ of order 2 which is not zero, while the

determinant of order 3 is zero. A second definition is that the rank of a square matrix is the number of independent column vectors, or the number of independent row vectors. Since the row rank equals the column rank for any square matrix, its rank equals the number of independent vectors contained in the matrix. But for a non-square matrix A of n×m its rank cannot exceed the smaller of the two dimensions of A i.e. $r(A) \leq \min(n,m)$.

By the linear independence of a set of 3 column vectors (a_1, a_2, a_3) in (1) we mean that none of the three vectors can be expressed as a linear combination of the other two i.e. there does not exist any nonzero values of the scalars x_1, x_2, x_3 such that $\sum_{i=1}^{3} a_i x_i = a_1 x_1 + a_2 x_2 + a_3 x_3 = 0$.

Let $A = \begin{bmatrix} 1 & 2 & 3 \\ 3 & 6 & 9 \\ 2 & 4 & 6 \end{bmatrix}$, choose $x_1 = 1$, $x_2 = 1$, $x_3 = -1$

we get $\Sigma a_i x_i = 0$, hence the three column vectors are not linearly independent; also a_1, a_2 or a_2, a_3 or a_3, a_1 are not linearly independent since

$$\begin{pmatrix} 2 \\ 6 \\ 4 \end{pmatrix} = 2 \begin{pmatrix} 1 \\ 3 \\ 2 \end{pmatrix}, \quad \begin{pmatrix} 2 \\ 6 \\ 4 \end{pmatrix} = (\tfrac{2}{3}) \begin{pmatrix} 3 \\ 9 \\ 6 \end{pmatrix}, \quad \begin{pmatrix} 3 \\ 9 \\ 6 \end{pmatrix} = 3 \begin{pmatrix} 1 \\ 3 \\ 2 \end{pmatrix}$$

hence its rank is one. This also follows from the determinantal test. The determinant of order 3 is zero since on expanding by the first row

$$|A| = 1(6 \times 6 - 4 \times 9) - 2(3 \times 6 - 2 \times 9) + 3(3 \times 4 - 2 \times 6) = 0.$$

Of order 2, there are nine subdeterminants, each having a zero value. Of order 1, there are nine subdeterminants i.e. the elements themselves, of which at least one is nonzero; hence $r(A) = 1$.

The concept of rank is useful in testing the solvability of a set of linear equations.

Test of Solvability

Consider the two equation system

$$x_1 + x_2 = 2$$
$$3x_1 + 3x_2 = 4$$

or,

$$\begin{bmatrix} 1 & 1 \\ 3 & 3 \end{bmatrix} \begin{pmatrix} x_1 \\ x_2 \end{pmatrix} = \begin{pmatrix} 2 \\ 4 \end{pmatrix}.$$

Rank of the coefficient matrix $A = \begin{bmatrix} 1 & 1 \\ 3 & 3 \end{bmatrix}$ is one and the rank of the augmented coefficient matrix $A(aug) = \begin{bmatrix} 1 & 1 & 2 \\ 3 & 3 & 4 \end{bmatrix}$ is two, since the subdeterminant formed by the last two columns is not zero. Hence

$$r(A) < r(A(aug)).$$

The equations are not mutually consistent i.e. they have no solution. In general, for any coefficient matrix A of m×n and the augmented coefficient matrix $A(aug) = [A:b]$ where $Ax = b$ is the system of m linear equations in n variables, we have no solution if

$$r(A:b) > r(A).$$

Although this rank test is useful, it may not be easy to apply it in practice for large systems. Hence in linear programming we apply another method for testing the solvability or consistency of a large linear system.

Unique Solutions

If $r(A) = r(A:b)$ i.e. the rank of the coefficient matrix equals the rank of the augmented coefficient matrix, then the m equations have at least one solution. In this case the vector b can be expressed as a linear combination of the independent columns of A i.e.

$$b = a_1 x_1 + a_2 x_2 + \ldots + a_k x_k$$

where $a_1, a_2, \ldots, a_k$ are the k independent column vectors of matrix A. Two cases arise

(i) If A is square and of full rank i.e. $r(A) = m$ and we have m linear equations $\sum_{j=1}^{m} a_{ij}x_j = b_i$, $i=1,2,\ldots,m$ then the solution vector x is unique (i.e. only one solution) and is given by $x = A^{-1}b$, A^{-1} being the inverse of A with the property that $AA^{-1} = I$, I being the identity matrix.

(ii) If A is not square, then there may be an infinite number of solutions. The most common example of this occurs in linear programming systems. For example consider the system

$$\begin{bmatrix} 1 & 2 & 7 \\ 3 & 3 & 6 \end{bmatrix} \begin{pmatrix} x_1 \\ x_2 \\ x_3 \end{pmatrix} = \begin{pmatrix} 4 \\ 3 \end{pmatrix}.$$

Here $r(A) = r(A:b) = 2$, since we have 2 rows and 3 columns of A and the first two rows of A define a nonzero determinantal value of -3. Moving terms involving x_3 to the right we obtain

$$\begin{bmatrix} 1 & 2 \\ 3 & 3 \end{bmatrix} \begin{pmatrix} x_1 \\ x_2 \end{pmatrix} = \begin{pmatrix} 4 - 7x_3 \\ x_2 - 6x_3 \end{pmatrix}.$$

For this system $A^{-1} = \begin{bmatrix} -1 & 2/3 \\ 1 & -1/3 \end{bmatrix}$, hence the solutions are

$x_1 = -2 + 3x_3$, $x_2 = 3 - 5x_3$.

But since any arbitrary value of x_3 may be chosen, we have infinite number of solutions.

Homogeneous System

If the right hand vector b is zero, we have a homogeneous system of m linear equations in n unknowns

$\sum_{j=1}^{n} a_{ij}x_j = 0$, $i=1,2,\ldots,m$ i.e. $Ax = 0$.

If $m=n$ and $r(A) = m$, then we have the trivial solution $x = A^{-1} 0 = 0$ which is unique. But if $r(A) < m$, then we have infinite number of solutions, which can be characterized by the eigenvalues to be discussed next.

1.4.2 Eigenvalues and Eigenvectors

For any <u>square</u> matrix A if a scalar λ and a nonzero vector x (i.e. not all the elements of x could be zero) can be found such that

$$Ax = \lambda x, \quad x \neq 0 \tag{1}$$

then λ is called an eigenvalue (also called latent or characteristic root) and x the corresponding eigenvector (also called latent or characteristic vector). Since for any nonzero scalar k we have $k(Ax) = k(\lambda x)$ and hence $A(kx) = \lambda(kx)$, therefore kx is also an eigenvector i.e. eigenvectors are determined upto a scalar multiple.

Write the system (1) as: $Bx = 0$, $B = (A - \lambda I)$, I = identity matrix of B is singular, then there exists a nonzero vector x satisfying (1). The singularity of B implies that the determinant $|B|$ of $|B|$ is zero. This determinantal equation $|B| = |A - \lambda I| = 0$ determines the eigenvalues e.g.,

Let $A = \begin{bmatrix} 1 & 5 \\ 2 & 4 \end{bmatrix}$, then $|A - \lambda I| = \begin{vmatrix} 1-\lambda & 5 \\ 2 & 4-\lambda \end{vmatrix}$.

This yields $\lambda^2 - 5\lambda - 6 = (\lambda + 1)(\lambda - 6) = 0$, i.e. $\lambda = -1, 6$. For the first eigenvalue $\lambda = \lambda_1 = -1$ we have the eigenvector from

$$\begin{bmatrix} 1 & 5 \\ 2 & 4 \end{bmatrix} \begin{pmatrix} x_1 \\ x_2 \end{pmatrix} = \begin{pmatrix} -x_1 \\ -x_2 \end{pmatrix}$$

or, $\begin{pmatrix} x_1 \\ -\frac{2}{5} x_1 \end{pmatrix}$ is the eigenvector where any arbitrary value of x_1 can be chosen.

For the second engenvalue $\lambda = \lambda_2 = 6$ we get similarly the eigenvector $\begin{pmatrix} x_1 \\ x_2 \end{pmatrix}$ where any value of x_1 can be chosen. In most common economic applications eigenvalues and eigenvectors are used in three different ways: (a) in solving linear difference and differential equations, (b) in testing the positive definiteness or negative definiteness of the Hessian matrix for determining the minimum or maximum of a quadratic or other nonlinear function, and (c) in computing maximal nonnegative eigenvalues for a class of matrices known as the Frobenius matrix, that is used to study the dynamic implications of the Leontief input-output model.

The following results hold for all symmetric matrices A which is square i.e. the element a_{ij} equals a_{ji}.

INTRODUCTION TO MATHEMATICAL ECONOMICS

<u>Theorem 1</u>. A symmetric matrix A has all eigenvalues real and it is positive (negative) definite if and only if all its eigenvalues are positive (negative). It is positive (negative) semidefinite if all its eigenvalues are nonnegative (nonpositive).

Alternatively, $A = (a_{ij})$ is positive definite if and only if the following subdeterminants along the main diagonal are all positive.

(a) $a_{11} > 0$; (b) $\begin{vmatrix} a_{11} & a_{12} \\ a_{21} & a_{22} \end{vmatrix} > 0$ and so on up to the n-th order value $|A| > 0$,

if it is of order n. For semidefiniteness some zero subdeterminants are allowed. But A is negative definite if and only if

(a) $a_{11} < 0$, (b) $\begin{vmatrix} a_{11} & a_{12} \\ a_{21} & a_{22} \end{vmatrix} > 0$, (c) $\begin{vmatrix} a_{11} & a_{12} & a_{13} \\ a_{21} & a_{22} & a_{23} \\ a_{31} & a_{32} & a_{33} \end{vmatrix} < 0$

and so on alternating in sign upto the n-th order. Again for semidefiniteness some zero determinants are allowed.

Let $A = \begin{bmatrix} 1 & -1 & -2 \\ 1 & 2 & 3 \\ -2 & 3 & 0 \end{bmatrix}$ then (a) $a_{11} = 1 > 0$

(b) $\begin{vmatrix} 1 & -1 \\ 1 & 2 \end{vmatrix} = 3 > 0$

(c) $|A| = -17 < 0$

hence A is neither positive definite, nor negative definite.

Let $A = \begin{bmatrix} 1 & -1 & 0 \\ -1 & 3 & 1 \\ 0 & 1 & 1 \end{bmatrix}$ then (a) $a_{11} = 1 > 0$

(b) $\begin{vmatrix} 1 & -1 \\ -1 & 3 \end{vmatrix} = 2 > 0$

(c) $|A| = 1 > 0$

it is positive definite. Its eigenvalues are all positive.

1.4.3 Frobenius and Metzler-Leontief Matrices

A Frobenius matrix is a square matrix $A = (a_{ij})$ of order n with all nonnegative elements i.e. $a_{ij} \geq 0$, all $i,j = 1,2,\ldots,n$. These matrices satisfy three most important conditions proved by the Frobenius Theorem:

(a) The matrix A has a nonnegative eigenvalue λ^* which is greater than or equal to the absolute value $|w|$ of any other eigenvalue i.e. $\lambda^* \geq |w|$, $\lambda^* \geq 0$.

(b) One can associate a nonnegative eigenvector $x^* \geq 0$ with the eigenvalue λ^*, which is also called the maximum eigenvalue or the Frobenius root, and,

(c) If any element a_{ij} of A increases, then λ^* stays the same or increases i.e. $\partial \lambda^* / \partial a_{ij} \geq 0$.

Note that under some additional assumptions (i.e. indecomposability) on the matrix A, the Frobenius root λ^* can be shown to be strictly positive.

One important use of the Frobenius Theorem is in computing the optimal growth path of a steady-state input-output model known as the von Neumann growth path. Let x be the n-tuple output vector and $Ax = \sum_{j=1}^{n} a_{ij}x_j$ be the inputs (raw materials) required to produce the output, where the matrix A has nonnegative elements and it is assumed to satisfy the condition of indecomposability. Let $r = 1+\alpha$, where α is the percentage markup on input costs, be the scalar rate of profit. We then ask does there exist a positive solution of the following model

$$\max r \text{ subject to } x \geq r Ax, \, x \geq 0.$$

If such a solution denoted by r^* exists, then one could construct the following eigenvalue problem

$$(A - gI) x = 0, \, g = 1/r$$

where $g^* = 1/r^*$ is the Frobenius eigenvalue and $x = x^*$ is the associated eigenvector. The Frobenius root g^* specifies a balanced growth path in some versions of Leontief's dynamic input-output model.

A second important square matrix in economics is the Metzler-Leontief matrix B with nonnegative diagonal elements $b_{ii} \geq 0$, i=1,2,...,n and nonpositive off-diagonal elements $b_{ij} \leq 0$, $i \neq j$. These matrices are easily related to a nonnegative Frobenius matrix by the relation $T = sI - B$ where the scalar number s is sufficiently large to make the matrix $T = (t_{ij})$ nonnegative i.e. $t_{ij} \geq 0$, all i and j. An open-static form of the Leontief input-output model has the following output balance equations in equilibrium

$$x_i = \sum_{j=1}^{n} a_{ij}x_j + c_i \, ; \quad i,j=1,2,\ldots,n$$

i.e. $x = Ax + c$, where $a_{ij} \geq 0$, all i,j

which states that the gross output for each sector i, (x_i) equals the sum of intermediate demand (Ax) and the final demand (c) by households and government. Suppose we predict the demand vector three years ahead as c_F, where c_F has all positive elements and assume that the elements of matrix A being input-output ratios did not change. Can we predict the equilibrium supply of output $\hat{x}$ by $\hat{x} = (I - A)^{-1} c_F$ where $\hat{x} \geq 0$ has all nonnegative elements? If the answer is yes, then these equilibrium nonnegative demand supply vectors $(c_F, \hat{x}_F)$ would maintain the relative prices of the base year. By defining $B = I-A$, we note that $b_{ii} > 0$ and $b_{ij} \leq 0$, $i \neq j$ by the two assumptions on the Leontief input-output matrix $A = (a_{ij})$ i.e. (1) $0 \leq a_{ij} \leq 1$ and (2) the A matrix is indecomposable in the sense that its two way interdependence cannot be reduced to one way interdependence i.e. for n=2 this means A cannot be reduced to any of the following:

$$\begin{bmatrix} a_{11} & 0 \\ a_{21} & a_{22} \end{bmatrix}, \quad \begin{bmatrix} a_{11} & a_{12} \\ 0 & a_{22} \end{bmatrix}, \quad \begin{bmatrix} a_{11} & 0 \\ 0 & a_{22} \end{bmatrix}.$$

Metzler-Leontief Theorem

Under these two conditions the Metzler-Leontief matrix B has the following three important properties:

(1) For any $c_F \geq 0$, there exists a nonnegative vector $\hat{x}$ such that $B\hat{x} = c_F$.

(2) The matrix $B = (I-A)$ is nonsingular so that we can solve for $\hat{x} = B^{-1} c_F = (I-A)^{-1} c_F$ giving us the equilibrium supply vector for any predicted nonnegative final demand, and

(3) The real parts of all the eigenvalues of B are all positive.

This theorem is useful also in studying the stability of the steady-state system corresponding to a lagged Leontief model e.g.

$$x(t) = Ax(t-1) + c$$

where $t=1,2,\ldots$ is time and each sector is assumed to have one unit lag in the input-output relation. The steady-state is defined by a value x^* where all $x(t)$'s for varying t have settled down. The steady-state vector x^* is defined for the above by

$$x^* = Ax^* + c, \quad \text{i.e.} \quad x^* = (I-A)^{-1} c.$$

It turns out by the Metzler-Leontief theorem that $x(t)$ tends to x^* as time $t \to \infty$ and all along the way $x(t)$ stays nonnegative since $(I-A)^{-1} \geq 0$ is satisfied.

PROBLEMS ONE

1. A function $y = f(x) = f(x_1, x_2, \ldots, x_n)$ is called homogeneous of degree r if $f(kx) = k^r f(x)$ for all scalars. Thus the production function $y = A x_1^a x_2^b$ with $a+b = 1$ with one output and two inputs x_1, x_2 is homogeneous of degree 1, since $f(kx) = A(kx_1)^a (kx_2)^b = k^{a+b}(Ax_1^a x_2^b) = ky$. Show that if each input is paid its marginal product $(\partial y / \partial x_i)$, then total factor payments exactly equals the total output.

2. (Euler's Theorem) For any homogeneous function of degree r, prove that total factor payments according to the marginal product rule exhausts the total product i.e.
$$\frac{\partial y}{\partial x_1} x_1 + \frac{\partial y}{\partial x_2} x_2 + \ldots + \frac{\partial y}{\partial x_n} x_n = ry.$$

3. Show that the following production function, known as the CES (constant elasticity of substitution) function
$$y = f(x_1, x_1) = [ax_1^{-c} + bx_2^{-c}]^{-1/c} \qquad a, b > 0$$
is homogeneous of degree one, where a, b, c are constants.

4. Test if the following functions are homogeneous of degree zero or not
$$\text{(a)} \ y = (x_1 + 2x_2)/(3x_1 + 4x_2), \qquad \text{(b)} \ y = \frac{x_1 + x_2}{(x_1^2 + x_2^2)^{1/2}}.$$

5. Consumers maximize their welfare $W = U(x) - px$, where $U(x) = a + bx - cx^2$ is utility derived from a composite good x, $\theta = (a,b,c)$ is positive and p is the market price given. Producers maximize their profit $\pi = px - C(x)$, where $C(x) = \alpha + \beta x + \gamma x^2$ is the total cost function, with $\alpha, \beta, \gamma > 0$. Derive the equilibrium price and quantity in the market if it is competitive.

6. In a macromodel the goods market is given by

$C_d = 0.95Y$ (consumption demand)
$I_d = 12 - 0.2r$ (investment demand)
Y_s: national income (real), r: rate of interest
$Y_s = C_d + I_d$

and the money market is given by

$$M_d = 0.25Y - 0.05r \quad \text{(money demand)}$$
$$M_s = 100 \quad \text{(money supply)}$$
$$M_d = M_s$$

(a) What is the equilibrium value of r and Y?
(b) How would you introduce the labor market in this system?
(c) If I_d is fixed at I_0 and r is fixed at r_0 then what is the value of the multiplier defined by $\Delta Y/\Delta I_0 = dY/dI_0$?

7. Solve the following:

(a) The marginal cost function is $MC = 10x + 20$, what is the total cost function?

(b) If the total cost function $T(x)$ is given by

$$T(x) = \begin{cases} 5 + 2x & \text{for } 0 \leq x \leq 2 \\ 10 + 2.5x & \text{for all } x > 2 \end{cases}$$

what is the marginal cost function?

8. Given the demand and supply equations where x = quantity and p = price

$$x_d = a - bp$$
$$x_s = c + dp$$
$$x_d = x_s = x$$

how do you know that there is no equilibrium solution? or, there are an infinite number of solutions?

9. An investor invests proportions x_1, x_2 in two stocks with average returns r_1, r_2. He maximizes $R = r_1 x_1 + r_2 x_2$ subject to $x_1 + x_2 = 1$. What would be the optimal solution if (a) $r_1 > r_2 > 0$, (b) $r_1 = r_2$?

10. (a) For any demand function $x = f(p)$ for a single good with x = quantity demanded and p = unit price, the price elasticity of demand, ε_p is defined as

$$\varepsilon_p = \frac{dx/x}{dp/p} = \frac{\% \text{ change in quantity}}{\% \text{ change in price}} = \frac{dx}{dp} \cdot \frac{p}{x}$$

show that ε_p can also be written as $\varepsilon = \dfrac{d \log x}{d \log p}$

[Hint: apply chain rule].

If the quantity of one good is influenced by prices of other commodities i.e. $x = f(p_1, p_2)$ then the partial elasticities of demand are

$$\varepsilon_{p_i} = \frac{\partial x}{\partial p_i} \cdot \frac{p_i}{x} \qquad (i=1,2).$$

(b) Henry Schultz became famous for his estimates of demand functions for a wide range of agricultural commodities. For the period 1915-19 he found the demand function for corn to be

$$p = 6570{,}000/q^{1.3}; \quad p = \text{price}, \; q = \text{demand quantity}.$$

What is the price elasticity (ε_p) of demand at the point when $q = 300{,}000$ bushels of corn are produced. Could farmers increase their total revenues from the sale of corn by increasing the price?

(c) Find the demand function $q = f(p)$ if the price elasticity ε_p of demand is $\varepsilon_p = -k$, k being a positive constant. What is the demand function if $k = 1.0$?

CHAPTER TWO
STATIC OPTIMIZATION AND ECONOMIC THEORY

Optimization occupies a most central place in modern economics. This is due to several reasons. As T.C. Koopmans noted in his 1975 Nobel memorial lecture: "The economist as such does not advocate criteria of optimality. He may invent them. ...the ultimate choice is made by the procedure of decision making inherent in the institutions, laws and customs of society." The distinction between positivistic and normative economic behavior is often useful for analytic convenience and many positive behavior may be derived by a process of implicit or subjective optimization. For example the shadow price of a constrained resource may be more realistic in reflecting its scarcity, though it may not always be realized in the actual market price. Secondly, the concept of an optimum may sometimes be replicated by a market equilibrium as in Adam Smith's theory of "the invisible hand" of a competitive system and one may ask if the price adjustments in the market tend to be equilibriating or not. Even in game theory models, where different players may pursue different and at times conflicting goals, the concepts of cooperative or noncoopertive equilibrium also involve optimization procedures. Recently the models of rational expectations in macroeconomic theory have emphasized specific normative hypotheses which assume that the agents use optimizing decisions in their expectation formation. Lastly, in dynamic situations over time disequilibrium models are increasingly applied in modern empirical economics which involve sequential search towards the dynamic optimum, or various adjustment costs due to errors.

2.1 Linear Programming

As a case of constrained optimization, linear programming (LP) offers perhaps the most applied tool in economics, management science and operations research. Due to linearity, the ordinary calculus method of locating the minimum or the maximum by applying the first derivative rule fails here, although by suitable use of the Lagrange multipliers one could still apply the first order condition with some modification. These modifications are spelled out by the Kuhn-Tucker Theorem which is applicable to a large class of nonlinear programming models known as concave programming.

As a simple example consider a farmer producing two crops x_1 (corn) and x_2 (wheat) with his available resources land (80 units = b_1) and labor (90 units = b_2). The input requirements are

$$a_{11} x_1 + a_{12} x_2 \leq b_1 \quad \text{(land)} \tag{1.1}$$
$$(1.0) \quad\quad (2.0) \quad\quad (80)$$

$$a_{21} x_2 + a_{22} x_2 \leq b_2 \quad \text{(labor)} \tag{1.2}$$
$$(3.0) \quad\quad (2.0) \quad\quad (90)$$

$$x_1 \geq 0, \; x_2 \geq 0. \tag{1.3}$$

The net return for each crop j is $c_j = p_j - v_j$, where p_j is the selling price and v_j is the unit variable cost. If $c_1 = 2.5$ and $c_2 = 2.0$, then the farmer's optimization model is

$$\text{Maximize } z = c_1 x_1 + c_2 x_2 \tag{1.4}$$
$$(2.5) \quad (2.0)$$

subject to (1.1), (1.2) and (1.3).

Note three features of this LP model. If the land constraint (1.1) were an equality, then it would have indicated an iso-land line i.e. various combinations of the two outputs requiring the same amount of land e.g. $x_1 = 0$, $x_2 = 40$ and $x_1 = 80$, $x_2 = 0$ are two such combinations. Similarly, for the labor constraint. However, the inequality sign in (1.1) says that some land may be left unused i.e. $b_1 = 80$ is the capacity level. One reason why it may be unused even at the optimal solution is that there may be mismatch with the availability of the second resource i.e. too much available land with too little available labor. Secondly, the model can easily handle more than two outputs and two semifixed inputs land and labor; also the production function is implicit here. Assume that we have the constraints as equalities (i.e. all resource capacities fully utilized) and then by inverting the coefficient matrix we obtain

$$\begin{pmatrix} x_1 \\ x_2 \end{pmatrix} = \begin{bmatrix} a_{11} & a_{12} \\ a_{21} & a_{22} \end{bmatrix}^{-1} = \begin{pmatrix} b_1 \\ b_2 \end{pmatrix}$$

or,
$$x = A^{-1} b = f(b).$$

If the vector x has all positive elements, then we obtain outputs as a linear combination of inputs, provided the coefficient matrix is nonsingular. This gives the production function. Thirdly, one cannot apply the first-order condition of calculus to maximize e.g., the linear function (1.4) i.e. $\partial z/\partial x_1 = c_1 \neq 0$, $\partial z/\partial x_2 = c_2 \neq 0$. However we can use Lagrange multipliers y_1 and y_2 for the two inequality constraints (1.1), (1.2) and form the Lagrangean $L(x,y)$:

$$L(x,y) = \sum_{j=1}^{2} c_j x_j + \sum_{i=1}^{2} y_i (b_i - \sum_{j=1}^{2} a_{ij} x_j) \tag{1.5}$$

and then apply a modified first-order condition. Note also that if the semifixed resources b_i can be freely bought and sold at a positive market price q_i, then the amounts of two inputs b_1, b_2 may also be optimally chosen along with the two outputs. The objective function in this case is profit π

$$\pi = \sum_{j=1}^{2} (p_j - v_j) x_j - \sum_{i=1}^{2} q_i b_i$$

and the new Lagrangean is

$$L(x,b,y) = \sum_{j=1}^{2} c_j x_j - \sum_{i=1}^{2} q_i b_i + \sum_{i=1}^{2} y_i (b_i - \sum_{j=1}^{2} a_{ij} x_j). \tag{1.6}$$

According to the Kuhn-Tucker Theorem, if the LP model has at least one solution satisfying all the constraints, then it has an optimal solution. If the optimal solution is denoted by the vector $x^{*T} = (x_1^*, x_2^*)$, then there must exist an optimal nonnegative vector $y^{*T} = (y_1^*, y_2^*)$ of Lagrange multipliers, jointly satisfying the following two necessary conditions

(1) $(\partial L/\partial x_j)_{x_j^*} = 0$, for $x_j^* > 0$, $j=1,2$
$\phantom{(\partial L/\partial x_j)_{x_j^*}} \leq 0$, for $x_j^* = 0$, $j=1,2$

(2) $(\frac{\partial L}{\partial y_i})_{y_i^*}$ $= 0$ for $y_i^* > 0$ $i=1,2$
$\phantom{(\frac{\partial L}{\partial y_i})_{y_i^*}} \geq 0$ for $y_i^* = 0$ $i=1,2$.

On applying these conditions to the Lagrangean (1.5) one obtains:

$$c_j - \sum_{i=1}^{2} y_i^* a_{ij} \leq 0 \quad \text{for } x_j^* \geq 0, \quad j=1,2$$

$$b_i - \sum_{j=1}^{2} a_{ij} x_j^* \geq 0 \quad \text{for } y_i^* \geq 0, \quad i=1,2.$$

The first necessary condition has two parts. One is that for x_j^* positive, the first order condition $\partial L/\partial x_j = 0$ of calculus holds. The second part says that if $\partial L/\partial x_j$

is negative at the optimal point, then the optimal output level x_j^* is zero. This is quite natural, since the Lagrangean L viewed as a function of output x is a net return or profit function and therefore $\partial L/\partial x_j$ is marginal return or profit. If marginal return or profit is negative for each unit produced at the optimal point, then it is generating a positive loss, hence $x_j^* = 0$. Likewise the L function viewed as a function of y (i.e. the imputed price of each resource) is a cost function which must have attained its minimum at the optimal point (y_1^*, y_2^*). But $\partial L/\partial y_i > 0$ implies $b_i > \sum_{j=1}^{2} a_{ij} x_j^*$, i=1,2 at the optimal point x^* i.e. resources not fully used, although the optimum has been reached, in that case $y_i^* = 0$ i.e. if land is surplus at the optimum its optimal Lagrange multiplier value is zero. Hence y_1^* is also called the imputed or the shadow price of land.

In the second part of each necessary condition we use an inequality sign

$$\left(\frac{\partial L}{\partial x_j}\right)_{x_j^*} \leq 0, \quad \left(\frac{\partial L}{\partial y_i}\right)_{y_i^*} \geq 0$$ because some degeneracy or redundancy may be present

in the constraints. But this is a technical computational point - not of much practical value.

The sufficiency condition for the pair (x^*, y^*) to be optimal is that the $L(x,y)$ function must be concave in elements of vector x for fixed $y = y^*$ and convex in the elements of vector y for fixed $x = x^*$. But since any linear function is both convex and concave, this sufficiency condition is always fulfilled for LP models.

2.1.1 Simplex Algorithms

The main reason for the wide popularity of LP models of optimization is the availability of a wide variety of computer routines, known as the simplex algorithm which has several variants such as revised simplex, parametric simplex and so on. The algorithm proceeds by jumping from one vertex of the convex feasible space to another by increasing the value of the objective at each step. Recently Karmarkar has proposed a method which combines interior movements in the feasible space with those along the edges.

The basic steps of the simplex method, closely related to the Kuhn-Tucker Theorem are very simple. The first step is to start with an initial feasible solution, then change it by replacing one activity (or output) at a time and finally there is a stopping rule when the optimal feasible solution is located. For our farming example we write the LP model as

$$\max z = 2.5x_1 + 2x_2 + 0x_3 + 0x_4$$
$$\text{subject to } x_1 + 2x_2 + x_3 = 80$$
$$3x_1 + 2x_2 + x_4 = 90 \quad (2.1)$$
$$\text{all } x_j \geq 0, \; j=1,2,3,4$$

where x_3, x_4 are nonnegative slack variables added on the left hand side of the constraints to convert the inequalities to equalities. Slack variables can be interpreted either as dummy outputs (activities) or unused inputs (resources). As unused resources they do not result in any real output, hence their net returns per unit are zero i.e. $c_3 = c_4 = 0$. The constraint equations in the LP model (2.1) now have 4 unknowns (x_1 through x_4) in two equations. Any 2 out of 4 can be selected in 6 ways and any one of these six selections is called a <u>basic solution</u> (or a vertex) and further if it satisfies the condition of nonnegativity of the x's, then it is called a basic feasible solution (or a feasible vertex, feasible in terms of <u>all</u> the constraints).

The slack variables x_3, x_4 have another role in testing the consistency of the two equations in (2.1). Since $x_3 = 80$, $x_4 = 90$ it provides the first basic feasible solution. In this sense the slack variables replace the rank rule of testing the consistency. However if one of the resources is negative, we may have to introduce another set of variables, called artificial variables, which along with the slack variables, provide the initial basic feasible solution if any.

The first step of the simplex routine for the above problem selects the initial basic feasible solution from the slack variables only and expresses in a tableau form the basic variables (x_3, x_4) in terms of the nonbasic variables (x_1, x_2) as follows

Tableau 1

Basic	Constant	Non-basic activities		Col. 1/Col. x_1
	1	x_1↓	x_2	$= r_1$
$x_3 =$	80	-1	-2	80/1 = 80
$x_4 =$	90	(-3)	-2	90/3 = 30 →
$z =$	0	2.5	2	

Thus row 1 reads $x_3 = 80 - x_1 - 2x_2$ and row 3 for profit reads $z = 0 + 2.5x_1 + 2x_2$. The marginal (potential) profitability of each of the nonbasic actiivty is

$$\Delta_1 = \partial z/\partial x_1 = 2.5 \text{ and } \Delta_2 = \partial z/\partial x_2 = 2.0.$$

In the next iteration one of these two nonbasic activities should enter the basis, because it would improve the profits. Note that so long as the third row has any positive marginal profitability, the necessary condition of optimality i.e. $(\partial L/\partial x_j)_* \leq 0$, j not in the basis is not fulfilled and hence the optimum is not reached. Which of the two non-basic activities, both having positive marginal profitability we should choose to bring in? The rule is: that nonbasic activity with the highest positive marginal profitability enters the new basis. Since $\Delta_1 >$

$\Delta_2 > 0$, therefore the new incoming activity is x_1. To find out the activity or variable in the current basis to be replaced i.e. the outgoing activity, we form the ratio r_1 of column marked 1 to column x_1 of the incoming activity, element by element and its absolute value is shown in the last column of Tableau 1. Column 1 here indicates the current value of the solution in terms of the basic variables and the value of the objective function, whereas column x_1 also called the <u>pivot column</u> indicates the input coefficients for the incoming activity. The minimum of the last column i.e. min(80,30) determines the outgoing basic activity which is obviously x_4 here. The coefficient (-3) in the intersection of row x_4 (outgoing activity) and column x_1 (incoming activity) is called the <u>pivot</u>.

In the second step we replace x_4 by x_1 as indicated by the respective arrows in Tableau 1. To complete this interchange we solve first for x_1 from the pivot row (row x_4) equation:

$$x_4 = 90 - 3x_1 - 2x_2$$

hence, $x_1 = \frac{1}{3}(90 - x_4 - 2x_2) = 30 - (1/3)x_4 - (2/3)x_2$

and substitute this new value of x_1 into the two old equations i.e. row x_3 and row z, in order to get rid of x_4. Thus we get

$$\begin{aligned} x_3 &= 80 - x_1 - 2x_2 = 80 - (30 - (1/3)x_4 - (2/3)x_2) - 2x_2 \\ &= 50 + (1/3)x_4 - (4/3)x_2 \\ z &= 0 + 2.5x_1 + 2x_2 + 0 = 2.5(30 - (1/3)x_4 - (2/3)x_2) + 2x_2 \\ &= 75 - (5/6)x_4 + (1/3)x_2. \end{aligned}$$

The second tableau then appears as follows:

Tableau 2

Basic	1	x_4	x_2↓	$r_2 = \|\text{col } 1/\text{col } x_2\|$
$x_3 =$	50	1/3	(-4/3)	37.5 →
$x_1 =$	30	-1/3	-2/3	45
$z =$	75	-5/6	1/3	

From row z we have

$$\Delta_4 = \partial z/\partial x_4 = -5/6, \quad \Delta_2 = \partial z/\partial x_2 = 1/3.$$

The highest positive is given by Δ_2, implying that x_2 should be incoming in the next iteration. Thus the x_2 column is the <u>pivot column</u>. We divide column 1 of constants by column x_2, element by element, ignoring the sign and then select the minimum i.e. min(37.5, 45) is 37.5 which identifies x_3 as the outgoing activity. The rule for selecting the outgoing activity is often called minimum feasibility or the weakest link principle i.e. once it is decided that x_2 is the new incoming activity, we ask at what maximum positive level x_2 can enter without violating feasibility. Thus we go back to rows x_3 and x_1 and solve for that maximum positive value of x_2 which makes x_3 and x_1 zero. Thus

$$x_3 = 50 + (1/3)x_4 - (4/3)x_2 = 50 + 0 - (4/3)x_2$$
$$x_1 = 30 - (1/3)x_4 - (2/3)x_2 = 30 - 0 - (2/3)x_2$$

where we have set $x_4 = 0$ since it is nonbasic and not incoming. It is clear that $x_3 = 0$ if $x_2 = 50/(4/3) = 37.5$ and $x_1 = 0$ for $x_2 = 30/(2/3) = 45$. We take the minimum of 37.5 and 45 in order to retain feasibility. Thus x_3 is the weakest link of the two nonbasic variables x_3 and x_1, since it goes to zero first. Note that if we choose $x_2 > 37.5$, then x_3 becomes negative, which is not feasible.

The third step is to determine when to stop the iterations. For this we look at the row z i.e. the last row of the tableau and test the sign of

$$\Delta_j = \partial z/\partial x_j, \; j = \text{nonbasic activity}.$$

If for <u>all</u> the <u>nonbasic activities</u>, Δ_j is negative or zero, then we have reached the optimal solution with maximum profits. If $\Delta_j < 0$ for all nonbasic j, then the optimal solution vector x^* is unique i.e. only one optimum output-mix. But for $\Delta_j \leq 0$, j nonbasic there may be more than one output-mix which give identical maximum profits. From Tableau 2 we see that $\Delta_2 = \partial z/\partial x_2 = 1/3$ is positive and hence we have to iterate still further i.e. by interchanging x_3 and x_2 as indicated by the arrows in Tableau 2. Proceeding as before we obtain the final tableau as:

Basic	Constant	Non-basic	
	1	x_4	x_3
$x_2 =$	37.5	1/4	-3/4
$x_1 =$	5	-1/2	1/2
$z =$	87.5	-3/4	-1/4

The last row z has $\Delta_4 = \partial z/\partial x_4 = -3/4$, $\Delta_3 = \partial z/\partial x_3 = -1/4$ which are all negative. Hence we stop. The optimum solution read off from the constant column 1 is $x_2^* = 37.5$, $x_1^* = 5$ which is unique and it yields the maximum profit $z^* = 87.5$.

2.1.2 Computational Aspects of Simplex

For some problems, specially very large size LPs some computational difficulties may arise in implementing the simplex algorithm. Some of the most common problems are as follows:

(i) Tie for the incoming or outgoing variable

If two or more variables, either outcoming (index j) or incoming (index i) have equal values of either $\Delta_j > 0$, or $r_i > 0$ then there is a tie. Any one may be selected in this case, or a more sophisticated rule known as the lexicographic rule (where if one variable is selected in one tableau, it is not selected in a later tableau in case of tie) may be adopted. For the case of outgoing variables, this may sometimes be due to what is known as <u>degeneracy</u>, which means that one or more basic feasible solutions (x_i^o say) in the constant column 1 are zero. In this case $r_i = |x_i^o / m_j|$ becomes zero.

(ii) Artificial variable technique

The slack variables as a group providing an initial basic feasible solution may fail in two situations when we have either some b_i values are negative or some constraints are equalities. Thus in the problem above if $b_1 = -80$ instead of +80, the two slack variables give $x_3 = -80$, $x_4 = 90$ which is not feasible. Alternatively, suppose the original LP model is max $z = 2.5x_1 + 2x_2$ subject to $x_1 + 2x_2 = 80$ and $3x_1 + 2x_2 = 0$ with x_1, x_2 nonnegative. There is no obvious initial basic solution now. In these cases we introduce the artificial variables x_5, x_6 say (over and above the slack variables if the latter are used also) to enlarge the constraint set as

$$x_1 + 2x_2 + x_5 = 80$$
$$3x_1 + 2x_2 + x_6 = 90 \qquad x_5, x_6 \geq 0 \qquad (2.2)$$

but since this revised problem would reduce to the original problem if x_5 and x_6 are zero in the optimal solution, we therefore force them to be zero by defining a high negative net return coefficient as (-M), where M is a large positive number. The revised objective function becomes then $z = 2.5x_1 + 2x_2 - Mx_5 - Mx_6$. Thus the high M method may be used to force an activity in or out of the optimal basis i.e. choose +M as the return coefficient in the first case and -M in the second case.

Note that the artificial variable technique may also be used when the b_i values are negative. For example if the first inequality were $x_1 + 2x_2 \leq -80$, we would subtract the artificial variable x_5 as

$$x_1 + 2x_2 - x_5 = 80, \; x_5 \geq 0$$

with $x_5 = 80$ along with $x_4 = 90$ providing the initial basic feasible solution. Conceptually any constraint can be enlarged by introducing both slack and artificial variables, e.g.

$$x_1 + 2x_2 + x_3 - x_5 = b_1$$
$$3x_1 + 2x_2 + x_4 - x_6 = b_2$$
$$\text{all } x_j \geq 0, \ j=1,2,\ldots,6$$

where x_3, x_4 are slack and x_5, x_6 are artificial variables. Thus, if $b_1 < 0$ and $b_2 > 0$ we choose x_5 and x_4 as the initial basic feasible solution; only qualification is that both x_3 and x_5 or, x_4 and x_6 cannot jointly be in any basic feasible solution.

Ex 1.   Minimize $z = 5x_1 + 3x_2$
Subject to (s.t.) $2x_1 + 4x_2 \leq 12$ (2.3)
$2x_1 + 2x_2 = 10$
$5x_1 + 2x_2 \geq 10$ $x_1, x_2 \geq 0$.

Adding x_3 as the nonnegative slack variable and x_4, x_5 and x_6 as the nonnegative artificial variables we transform the model as

$$\text{Min } z = 5x_1 + 3x_2 + 0x_3 + Mx_4 + 0x_5 + Mx_6$$
$$\text{s.t. } 2x_1 + 4x_2 + x_3 = 12$$
$$2x_1 + 2x_2 + x_4 = 10$$
$$5x_1 + 2x_2 - x_5 + x_6 = 10$$
$$x_{ij} \geq 0, \ j=1,2,\ldots,6.$$

Clearly the initial basic feasible solution is $x_3 = 12$, $x_4 = 10$ and $x_6 = 10$ i.e. it is composed of one slack variable (x_3) and two artificial variables x_4, x_6:

Tableau 1

Basic	Constant	Nonbasic			r_1
	1	$x_1\downarrow$	x_2	x_5	
x_3	12	2	4	0	12/2 = 6
x_4	10	2	2	0	10/2 = 5
$\rightarrow x_6$	10	(5)	2	-1	10/5 = 2$\rightarrow$
z	20M	5-7M	3-4M	M	

Here $\Delta_1 = \partial z/\partial x_1 = 5-7M$, $\Delta_2 = 3-4M$ and $\Delta_3 = M$. Applying our variable entry criterion, noting that we are minimizing we select the smallest positive value of Δ_j i.e. x_1 is the entering basic variable, since for $M \gg 0$, $5-7M < 3-4M$. The current basic variable to be removed as x_1 becomes positive is chosen using the variable removal criterion i.e. the smallest absolute value of the ratio denoted by $r_1 = |\text{col } 1/\text{col } x_1|$, where col x_1 is the pivot column. This gives x_6 as the outgoing variable and the entry 5 at the intersection of row x_6 and column x_1 is the pivot element. The iterative operations proceed as before, till we reach the final tableau

		Nonbasic		
Basic	1	x_3	x_4	x_6
x_2	1	1/3	-1/2	0
x_5	12	-3/2	4	-1
x_1	4	-1/2	1	0
z		+1	+M-7/2	+M

Since the last row has all positive elements we have a unique minimum and we stop. In the maximization problem this row has all negative elements for a unique maximum.

(iii) **Modified simplex methods**

In problems known as the transportation type LP models, the coefficient matrix has a special structure, as each a_{ij} coefficient takes only two values either zero or one. Hence more efficient computational routines are available in place of the standard simplex routine of selecting the initial basic feasible solution. As an example of this problem assume that a homogeneous product is produced in two locations (i=1,2) and sold in 3 markets (j=1,2,3) where the transportation cost is c_{ij} per unit of x_{ij}, where x_{ij} is the amount produced in location i but sold in market j. Total demand is d_j in market j and total capacity output is k_i in factory located in i. Then the optimization model is to choose (x_{ij}) by minimizing total transportation cost C

$$\text{Min } C = \sum_{i=1}^{2} \sum_{j=1}^{3} c_{ij} x_{ij}$$

$$\text{s.t.} \quad \sum_{j=1}^{3} x_{ij} \leq k_i \quad i=1,2 \quad \text{(capacity constraint)}$$

$$\sum_{i=1}^{2} x_{ij} \geq d_j \quad j=1,2,3 \quad \text{(demand constraint)}$$

$$\text{all } x_{ij} \geq 0, \quad i=1,2; \; j=1,2,3.$$

STATIC OPTIMIZATION AND ECONOMIC THEORY

The objective function could easily be modified to include a linear cost of production.

2.1.3 Economic Applications
We consider some useful applications of LP models in micro and macroeconomics.

Ex 2 (Peak-load pricing)

Consider the long distance service of a telephone company, where there are two rates p_1, p_2 per minute of call with p_1 higher than p_2. The former rate is for weekdays when capacity is fully utilized and demand is at peak, whereas p_2 is for weekends when capacity is not fully utilized. How can one justify the two rates for the same service? Is it price discrimination? The answer is no. Since the marginal cost of using extra capacity is zero, when capacity is not fully utilized, therefore the off-peak price p_2 is lower than p_1, where p_1 has to include a positive marginal cost of capacity utilization.

The same problem arises for a competitive firm (or a firm regulated by public utility commission) selling electricity at a unit price p during 24 hours, of which some are peak (i.e. daytime in weekdays when capacity of the plant is fully utilized) and some off-peak (when capacity is not fully utilized). With x_i as the units of power supplied in hour i and $C = C(x_1, x_2, \ldots, x_{24})$ as the total cost function and k as plant capacity measured in output assumed to be fixed in the short run, the optimal decision problem for the firm is to

$$\text{Max } z \text{ (profits)} = \sum_{i=1}^{24} px_i - C(x_1, \ldots, x_{24})$$

$$\text{s.t.} \quad x_i \leq k \qquad\qquad\qquad (3.1)$$
$$\quad x_i \geq 0; \quad i=1,2,\ldots,24.$$

On defining the Lagrangean function $L(x,y)$ where $y = (y_i)$ is the vector of Lagrange multipliers

$$L(x,y) = \sum_{i=1}^{24} px_i - C(x_1, \ldots, x_{24}) + \sum_{i=1}^{24} y_i(k-x_i)$$

the necessary conditions for an optimal feasible solution are by Kuhn-Tucker theory

$$\partial L(x,y)/\partial x_i = p_i - MC(x_i) - y_i \leq 0, \quad x_i \geq 0$$
$$\partial L(x,y)/\partial y_i = k - x_i \geq 0; \, y_i \geq 0; \quad \text{all } i=1,\ldots,24. \qquad (3.2)$$

If the cost function $C = C(x)$ is convex or linear, these necessary conditions are also sufficient for obtaining maximal profits. Since x_i is positive for all practical purposes, then the first condition of (3.2) implies at the optimal point:

$$p = MC(x_i) + y_i, \quad MC(x_i) = \frac{\partial C(x)}{\partial x_i} = \text{marginal cost}.$$

But by the other condition of (3.2)

and
$$y_i > 0, \text{ if } x_i = k \text{ (peak demand; capacity fully utilized)}$$
$$y_i = 0, \text{ if } x_i < k \text{ (off peak; unused capacity)}.$$

In the first case, $p > MC(x_i)$ and in the second $p = MC(x_i)$ i.e. price equals marginal cost, since $y_i = 0$. Thus we have a two-part pricing rule which satisfies the criterion of Pareto-optimality i.e. some gain somewhere without any loss anywhere. This is because the consumers gain by switching their demand to off-peak periods when the prices are lower. If one price (i.e. the average of the peak and off-peak) prevailed by regulation for instance, some consumers may not have called at all; so the society gains in terms of higher output induced by higher demand. The supplier gains also, since his marginal cost is covered in both periods.

<u>Ex 3</u> (Leontief's input-output model)

Consider the static input-output (IO) model of Leontief, where the gross output x_i of each sector i (i=1,2,...,n) is demanded partly as raw materials in other sectors ($x_{ij} = a_{ij}x_j$; i,j=1,2,...,n) and partly as final demand (d_i) by households and other final users. It is assumed that there is only one primary input called labor which is proportional to output in each sector. The total labor requirement is therefore

$$L = \sum_{i=1}^{n} h_i x_i, \quad h_i = \text{constant positive labor coefficient}.$$

The households can obtain the various goods x_i by offering their labor services. Hence the optimal decision that the households must make is to satisfy their demand in the cheapest possible way. This leads to the IO LP model

$$\operatorname*{Min}_{x} L = \sum_{i=1}^{n} h_i x_i$$

s.t.
$$x_i - \sum_{j=1}^{n} a_{ij} x_j \geq d_i \qquad (3.3)$$

$$x_i \geq 0; \quad i,j=1,2,\ldots,n$$

where $R_i = \sum_{j=1}^{n} x_{ij} = \sum_{j=1}^{n} a_{ij}x_j$ is the total raw material demand by all other sectors j from sector i, with a_{ij} being the nonnegative IO coefficient $a_{ij} = x_{ij}/x_j$ and d_i is the positive level of households' demand. The a_{ij} coefficients of the Leontief's IO model have two important properties unlike that of any arbitrary LP model. One is that all the coefficients a_{ij} are nonnegative and they satisfy the so-called Hawkins-Simon conditions which guarantee that the n linear equations in x_i (i=1,...,n)

$$x_i - \sum_{j=1}^{n} a_{ij}x_j = d_i; \quad i=1,2,\ldots,n \tag{3.4}$$

have nonnegative (positive) solutions for any nonnegative (positive) demand d_i. These conditions require that all the princpal minors of the coefficient matrix of (3.4) are positive i.e.

$$1 - a_{11} > 0, \quad \begin{vmatrix} 1-a_{11} & -a_{12} \\ -a_{21} & 1-a_{22} \end{vmatrix} > 0, \ldots, \begin{vmatrix} 1-a_{11} & \cdots & -a_{1n} \\ \vdots & & \vdots \\ -a_{n1} & \cdots & 1-a_{nn} \end{vmatrix} > 0$$

This we have already discussed in Chapter One under Metzler-Leontief Theorem.

Two interpretations of the LP version of the IO model are very important. One is the duality and the second is the Substitution Theorem. Let $x^* = (x_i^*)$ be an n-tuple optimal solution vector for any fixed demand vector $d = (d_i) \geq 0$. Then by the Kuhn-Tucker Theorem the vector $p^* = (p_i^*)$ of optimal shadow prices must also solve the dual problem:

$$\max z = \sum_{i=1}^{n} p_i d_i = \sum_{j=1}^{n} p_j d_j$$

$$\text{s.t.} \quad p_j - \sum_{i=1}^{n} p_i a_{ij} \leq h_j \tag{3.5}$$

$$p_j \geq 0, \quad i,j=1,2,\ldots,n$$

and by strong duality theorem we must have

$$\sum p_j^* d_j = \sum h_i x_i^*$$

which gives the two interpretations of national income as the value of the sum of net sectoral outputs or the sum of net factor costs. Also, by Hawkins-Simon conditions the optimal (shadow) prices p_j^* are positive if h_j's are positive and hence at the optimal solution

$$p_j^* = \sum_{i=1}^{n} p_i^* a_{ij} + h_j, \quad j=1,2,\ldots,n$$

which is the familiar condition of competitive equilibrium: price equals marginal cost made up of marginal raw material cost and marginal labor cost.

Now suppose each good j is producible by several independent processes or techniques rather than one; hence each a_{ij} becomes $a_{ij}(u)$ where the argument $u=1,2,\ldots,k$ indicates that there are k processes for each output. We assume however that the new coefficients $a_{ij}(u)$ still satisfy the Hawkins-Simon condition and the condition of indecomposability. Now suppose the demand vector $d = (d_i)$ changes from its base year values, still remaining nonnegative however. Would it not change the optimal output-mix in the sense of one optimal set of $u_{ij}(1)$ for $u=1$ and another $u_{ij}(2)$ for $u=2$? The answer is no. This is due to the Substitution Theorem which says that if a particular set of input coefficients ($a_{ij}(1)$ say) is optimal for a fixed nonnegative demand vector d, it will remain so for any other nonnegative values of demand. A simple way to prove the theorem is to go back to the simplex criterion $\Delta_j = \partial z/\partial x_j$, j = nonbasic applied in simplex tableaus before and note that the stopping rule: $\{\Delta_j \leq 0$ for all j not in the basis$\}$ does not depend on the demand components d_i of the IO model except of course through feasibility which necessarily holds by the Hawkins-Simon condition.

<u>Ex 4</u> (goal programming model)

Consider a company producing two types of video tapes in quantities x_1 and x_2 using two limiting resources $b_1 = 80$ and $b_2 = 60$. The decision model used is

$$\begin{aligned}
\text{Max profit } z &= 100x_1 + 150x_2 \\
\text{s.t.} \quad 4x_1 + 2x_2 &\leq 80 \\
2x_1 + 2x_2 &\leq 60 \\
x_1 \geq 0, \; x_2 &\geq 0.
\end{aligned} \quad (3.6)$$

The optimal solution to this problem is $x_1 = 0$ units and $x_2 = 30$ units with a maximum profit $z = \$4500$. Suppose now the company is planning to make a major capital investment to improve its manufacturing efficiency and hence it feels that it can no longer simply seek to maximize profit. However it would like to achieve some satisfactory or target level of profit during the period of capital investment. Let $5000 be the target level of profit or goal selected. Then define d_1^+ as the over-achievement of the target profit and d_1^- as the underachievement so that the profit goal constraint becomes

$$100x_1 + 150x_2 - (d_1^+ - d_1^-) = 5000 \quad \text{(profit goal constraint)}.$$

The entire goal programming model then becomes

Minimize d_1^-
s.t. $4x_1 + 2x_1 \le 80$
$2x_1 + 2x_2 \le 60$ (3.7)
$100x_1 + 150x_1 - d_1^+ + d_1^- = 5000;$ $x_1, x_2, d_1^+, d_1^- \ge 0.$

Note that the underachievement variable d_1^- here is the amount by which the actual profit may fail to reach the target profit and the overachievement variable d_1^+ is the amount by which the actual daily profit may exceed the target or goal profit. In this example the underachievement of the profit goal is considered undesirable and hence it is minimized in the objective function.

Suppose instead of the profit goal of $5000, the company intends to achieve a sales goal of at least 20 for each of x_1 and x_2. Then for the first sales goal: $x_1 = 20$ we define two variables d_1^+ for overachievement and d_1^- for underachievement and likewise d_2^+ and d_2^- for the second sales goal $x_2 = 20$. The goal programming model then becomes

Minimize $C = d_1^- + d_2^-$
s.t. $4x_1 + 2x_2 \le 80$
$2x_1 + 2x_2 \le 60$ $x_2 - d_2^+ + d_2^- = 20$
$x_1 - d_1^+ + d_1^- = 20$
$x_1, x_2, d_1^+, d_1^-, d_2^+, d_2^- \ge 0.$

Here we minimize the underachievement of the two sales goals, both getting equal weights. It is clear that unequal weights w_1, w_2 can be also assigned to the two goals (i.e $C = w_1 d_1^- + w_2 d_2^-$) and conflicting goals may also be handled by this formulation. The latter aspect is useful in econometric estimation when we apply the least absolute value method of estimation instead of least squares. Note two useful features of the goal programming model. The decision model need not have a preassigned objective function i.e. any goal or target would do. In this sense it defines a satisficing rather than optimizing approach, to use a concept developed by Herbert Simon. Second, nearness to a goal, even when the goal is unachievable due to conflict or to a very high value yields a surrogate objective function that generates an approximate solution. Thus even when the linear constraints of a regular LP model are mutually conflicting, so that there is no feasible solution, the goal programming model generates an approximate solution that is infeasible in the strict sense but it is also least infeasible (e.g. least underachievement of the sales goal).

Ex 5 (use of shadow prices)

Consider a farmer producing two crops x_1, x_2 using two resources: land $b_1 = 4000$ and the other composite resource $b_2 = 100$. The LP model is

$$\text{Max profit } z = 40x_1 + 70x_2$$
$$\text{s.t.} \quad 10x_1 + 50x_2 \leq 4000 \text{ (land)}$$
$$x_1 + x_2 \leq 100 \text{ (other resource)} \quad (3.7)$$
$$x_1, x_2 \geq 0.$$

The optimal solution is $x_1^* = 25$, $x_2^* = 75$, $z^* = \$6250$ and the two optimal shadow prices of land and the other resource are $y_1^* = 0.75$ and $y_2^* = 32.5$ respectively. Suppose the farmer now decides to increase his land resource by a small positive amount k. How much should he add in an optimal sense?

If the LP model (3.7) remains the same except that its first constraint now becomes

$$10x_1 + 50x_2 \leq 4000 + k$$

we may assume that the optimal shadow price of land y_1^* would still remain the same. Then we must have after k units are added:

$$z^* = 40x_1^* + 70x_2^* = 0.75(10x_1^* + 50x_2) + 32.5(x_1^* + x_2^*) \text{ (original solution)}$$

and

$$z^* \leq 0.75(4000 + k) + 3250 = 6250 + 0.75k.$$

Hence the extra profit will never exceed 0.75k. In fact if k = 1000 then the farmer can realize an additional profit of $750 by letting new values of x_1 and x_2 as $\bar{x}_1 = 25 - 0.025k = 0$ and $\bar{x}_2 = 75 - 0.025k = 50$ respectively.

2.1.4 Some Theory of LP

From an applied viewpoint two aspects of the theory of LP are very important. The first is the duality theory which is closely related to the Kuhn-Tucker (KT) Theorem on nonlinear programming. The second is the transformation of an LP model into an equivalent zero-sum two-person game.

Duality and Its Applications

The KT Theorem, when applied to LP models uses the Lagrangean function to define a set of necessary conditions (which are also sufficient) for the vectors x^* and y^* to solve the problem. First, it shows that if an optimal n-tuple vector x^* solves the LP model in primal form:

Primal

$$\text{Max } z = \sum_{j=1}^{n} c_j x_j \qquad \text{Max } z = c^T x$$

or,

$$\text{s.t. } \sum_{j=0}^{n} a_{ij} x_j \leq b_i \qquad \text{s.t. } Ax \leq b \qquad (3.8)$$

$$x_j \geq 0; \; i=1,2,\ldots,m \qquad x \geq 0$$
$$ j=1,2,\ldots,n$$

then there must exist an m-tuple optimal vector y^* of Lagrange multipliers $y = (y_i)$, $y \geq 0$ which minimizes the Lagrangean function

$$L(x^*,y) = c^T x^* + y^T (b - Ax^*) \qquad (3.9)$$

where the optimal (i.e. maximizing) value x^* of x has to be used. The necessary condition for x^* to be an optimal value i.e a maximizer is that

$$\left(\frac{\partial L(x,y)}{\partial x}\right)^* \leq 0, \; x^* \geq 0, \; y^* \geq 0$$

where the asterisk on the partial derivative means that it is evaluated at the optimal level. By evaluating this partial derivative at x^* we obtain

$$c^T - y^T A \leq 0$$

$$c_j - \sum_{j=1}^{m} y_i a_{ij} \leq 0 \qquad \text{or,} \qquad (3.10)$$

$$y_i \geq 0 \qquad\qquad y \geq 0$$

On taking only the terms involving y in the Lagrangean function (3.9) and using (3.10) as a condition for ensuring the maximizing value of x, we obtain the dual problem:

Dual $\quad \min_{y \in W} C = b^T y; \quad W = \{y \mid y^T A \geq c^T, \; y \geq 0\}$

or,

$$\text{Min } C = \sum_{i=1}^{m} b_i y_i$$

$$(3.11)$$

$$\text{s.t. } \sum_{i=1}^{m} y_i a_{ij} \geq c_j; \; y_i \geq 0; \; i=1,2,\ldots,m$$
$$\phantom{\text{s.t. } \sum_{i=1}^{m} y_i a_{ij} \geq c_j; \; y_i \geq 0;\;} j=1,2,\ldots n.$$

On comparing this dual form (3.11) with the primal (3.8), several features of difference become apparent. First, the structure of the two problems are different e.g. the primal has maximization as the objective, the dual has minimization; the inequalities $Ax \leq b$ are reversed as $y^T A \geq c^T$ (i.e. $A^T y \geq c$ on taking transpose of both sides). Second, by interpreting x_j as the Lagrange multiplier for each of the n dual constraints $\sum_{i=1}^{m} y_i a_{ij} \geq c_j$, one can easily derive the following condition, known as <u>weak duality</u>:

$$\sum_{j=1}^{n} c_j x_j \leq \sum_{j=1}^{n} \sum_{i=1}^{m} y_i a_{ij} x_j \leq \sum_{i=1}^{m} b_i y_i \qquad (3.12)$$

This is obtained by multiplying the primal constraint of (3.8) by $y_i \geq 0$ and summing over i and similarly by multiplying the dual constraint of (3.11) by $x_j \geq 0$ and summing over j. Weak duality implies that for any output vector x feasible in terms of the primal constraints and any shadow price vector y feasible in terms of the dual constraints, feasible profit is bounded by the total imputed cost. <u>Strong duality</u> says that if both x^* and y^* are both optimum feasible in their primal and dual LP models respectively, then we must have

$$z^* = \sum_{j=1}^{n} c_j x_j^* = \sum_{i=1}^{m} \sum_{j=1}^{n} y_i^* a_{ij} x_j^* = \sum_{i=1}^{m} b_i y_i^* = C^*$$

$$x_j^* \geq 0, \; y_i^* \geq 0, \; i=1,2,\ldots,m; \; j=1,2,\ldots,n.$$

Note that the inequality in case of <u>weak duality</u> (3.12) is now transformed to equality in case of <u>strong duality</u>. Thus (x^*,y^*) may be called a maximin pair, since x^* is a maximizer of the primal objective function and y^* is a minimizer for the dual. This leads to the saddle point inequality in terms of the Lagrangean function

$$L(x,y^*) \leq L(x^*,y^*) \leq L(x^*,y)$$

for all nonnegative vectors x, y, x^*, y^*. This may also be written as

$$\min_{y} L(x,y) \leq \max_{x} \min_{y} L(x,y) \leq \max_{x} L(x,y).$$

Note that we can derive from the strong duality relation the sensitivity of the optimal profit and optimal cost due to small variations in the different parameters

$$\partial C^*/\partial b_i = \partial z^*/\partial b_i = y_i^* \geq 0$$
$$\partial z^*/\partial c_j = x_j^* \geq 0, \; \partial z^*/\partial a_{ij} = -y_i^* x_j^*$$

The last result is useful in determining which input-output coefficient is most critical in altering the optimal profit level.

The necessary conditions for the pair (x^*, y^*) to be a maximin pair in the LP model are as follows according to the KT Theorem:

$$\left(\frac{\partial L(x,y)}{\partial x}\right)^* \leq 0; \qquad \left(x^T \frac{\partial L(x,y)}{\partial x}\right)^* = 0 \qquad (3.13)$$

$$x^* \geq 0, \ y^* \geq 0$$

and

$$\left(\frac{\partial L(x,y)}{\partial y}\right)^* \geq 0; \qquad \left(y^T \frac{\partial L(x,y)}{\partial y}\right)^* = 0 \qquad (3.14)$$

$$y^* \geq 0, \ x^* \geq 0.$$

These two results jointly are known as <u>complementary slackness conditions</u> which must hold for any optimal pair (x^*, y^*). They have the following economic interpretation, when the primal LP is viewed as a profit-maximizing model for a firm choosing an optimal output-mix with given resources and hence the dual as minimizing total resource costs. The first part of (3.13) evaluated at the j-th component of x says that if $x_j^* \geq 0$ is an optimal output belonging to x^*, then

$$(\partial L/\partial x_j)^* = c_j - \sum_{i=1}^{m} y_i^* a_{ij} \leq 0$$

i.e. net excess profits per unit of optimal output x_j^* is either zero or negative, when the total cost of resources used is evaluted at the optimal shadow prices $y_i^* \geq 0$. The second part of (3.13) says that

if $\quad c_j - \sum_{i=1}^{m} y_i^* a_{ij} < 0$, then $x_j^* = 0$ \hfill (3.15a)

and if $\quad c_j - \sum_{i=1}^{m} y_i^* a_{ij} = 0$, then $x_j^* \geq 0$ (for unique optimum, $x_j^* > 0$) \hfill (3.15b)

The relation (3.15a) gives the reason why x_j^* should not be produced at all, since it entails marginal loss. The familiar rule of price equals marginal cost which holds in competitive markets holds in case of (3.15b), where $x_j^* > 0$ holds for unique optimum and when there is no degeneracy in the problem.

Likewise for the dual variables in (3.14) we have the otpimality rules

if $\quad (\partial L/\partial y_i)^* = b_i - \sum_{j=1}^{n} a_{ij} x_j^* > 0$, then $y_i^* = 0$

(i.e. surplus resource at the optimum has a zero shadow price)

and

if $(\partial L/\partial y_i)^* = 0$, then $y_i^* \geq 0$ (for unique optimum $y_i^* > 0$).

These are the two rules of optimal shadow prices y_i^* i.e. an oversupplied resource must have a zero imputed price, whereas a fully utilized resource must have a positive marginal value $y_i^* = \partial z^*/\partial b_i > 0$ except for cases of degeneracy or nonunique optimum. Optimal shadow prices y_i^* therefore represent either the marginal cost of resources, or its equivalent, the marginal profitability.

Since the shadow prices y_i^* are most widely used in economic theory and practice, one may mention here some of its uses and limitations. First, we note that if resources b_i can be bought and sold in any amount in a competitive market at a positive price q_i, then it is always optimal to use that much of every resource b_i at which $y_i^* = q_i$. This is so because the case $y_i^* < q_i$ implies that marginal profitability is less than the marginal cost of buying the resource leading to a marginal loss; also $y_i^* > q_i$ implies that profits could still be increased by adding more resources. However if the resources are not perfectly divisible due to their lumpiness, one may have at the optimal level $y_i^* \leq q_i$ for some i. Secondly, the vector y^* of imputed prices y_i^* is computable even when the market prices q_i of resources are not available. Thus by varying the availability limit of b_i, a parametric variation of y_i^* may be evaluated resulting in a piecewise linear demand curve in the form $y_i^* = h(b_i)$ with $\partial y_i^*/\partial b_i \leq 0$. Thus for many public agencies like water boards, the impact of shortage of a resource (e.g. water) may thus be evaluated in terms of the increase in shadow prices due to resource shortage.

Another aspect of shadow pricing rule is that it locates hidden inefficiencies due to nonoptimal resource-mix or mismatch in resource combinations. Thus consider a farmer in a less developed country producing three crops x_1, x_2, x_3 with two inputs, land and labor and maximizing his net income z as follows:

$$\begin{aligned} \max z = \quad & x_1 + 3x_1 + 2x_3 \\ \text{s.t.} \quad & x_1 + 2x_2 + x_3 = 4 \text{ (land)} \\ & 3x_1 + x_2 + 2x_3 = b_2 \text{ (labor)} \\ & x_j \geq 0, \; j=1,2,3. \end{aligned} \qquad (3.16)$$

If labor supply were 9 units, the optimal solution would be $x_1^* = 1.0$, $x_2^* = 0$, $x_3^* = 3$ and $z^* = 7$; but if labor supply is increased to $b_2 = 12$ due to additional births in the family, the maximal net income becomes $z^* = 4$ with $x_1^* = 4$, $x_2^* = 0 = x_3^*$. The consequence of forcing more labor on the farm is to reduce the earlier level of maximal income, since it produced more distortion and not less. This is so, because due to equality constraints, the optimal shadow prices of labor can be of any sign and it became negative when $b_2 = 12$. Thus we note one important feature of the equality constraint in LP. If a constraint in the primal problem is an equality to start with i.e.

$$\sum_{j=1}^{n} a_{ij} x_j = b_i$$

then the shadow price y_i^* of b_i at the optimal solution can take on any sign: positive, negative or zero. Likewise if any dual constraint were an equality to start with

$$\sum_{i=1}^{m} y_i a_{ij} = c_j$$

then the shadow price x_j^* of c_j can be of any sign: positive, negative or zero. Note however that the shadow price y_i^* is a function of all the parameters (a_{ij}, b_i, c_j) and not simply the resource b_i, although in LP models this dependence is implicit and not explicit.

A fourth use of shadow prices is in models of national planning, where the central authority, usually the national planning board has to allocate some critical central resource like foreign exchange optimally between the sectors. When the markets do not exist in competitive form, or the market prices do not adequately reflect the true scarcity of resources like foreign exchange, the shadow prices may be used by the central planning authority to guide optimal allocation between sectors which are asked to compete for higher allocation through bidding at higher shadow prices i.e. higher shadow prices indicate higher marginal profitability. Thus the shadow price guided allocation, sometimes followed in French planning systems helps secure an optimal scheme of decentralization; this problem can also arise in a large corporation with divisional subsidiaries. This process of revising a set of interim shadow prices by competitive bidding between sectors till the optimal allocation pattern is reached is applied also in decomposition algorithms for large-scale LP models.

We must note however two limitations of the optimal solution of an LP model. One is that it allows restricted substitution i.e. the optimum solution is not fully diversified. This is due to the dimensional restriction of the optimal basis. Thus in the LP model (3.8), the optimal solution cannot have more than m positive x_j's, when m is less than or equal to n. For example in the problem

$$\max z = \sum_{j=1}^{n} c_j x_j$$

$$\text{s.t.} \sum_{j=1}^{n} x_j = 1, \quad x_j \geq 0$$

where c_j's are all positive, the optimal solution is $x_k^* = 1$ and all other x_j^*'s are zero, where k is indicated by $c_k = \max_{1 \leq j \leq n} c_j$. Thus only one output is positive at the optimal solution. But if the objective function were quadratic e.g.
$z = \sum_{j=1}^{n} c_j x_j - \frac{1}{2} \sum_{j=1}^{n} x_j^2$, then by using the Lagrange multiplier y_1 for the single

constraint we obtain the optimal solution

$$x_j^* = c_j - y_1^*, \quad j=1,2,\ldots,n$$

$$y_1^* = (\sum_{j=1}^{n} c_j - 1)/n. \tag{3.17}$$

All the n x_j's are positive at the optimal solution.

A second limitation, we have mentioned before is that the shadow prices of any resource depend implicitly on all the parameters (c, A, b). In a quadratic program this dependence is explicit. For instance the optimal shadow price in the quadratic program (3.17) depends explicitly on x_j^*'s:

$$y_1^* = c_j - x_j^*, \quad j=1,2,\ldots,n. \tag{3.18}$$

This is the derived demand function showing that as demand for output rises, the shadow price declines i.e. on summing both sides of (3.18) we obtain the total demand function

$$x^* = \bar{c} - n y_1^*, \quad x^* = \sum_{j=1}^{n} x_j^*, \quad \bar{c} = \sum_{j=1}^{n} c_j.$$

Unlike the positivistic demand function, this is a normative demand function which can be utilized for policy questions. For instance if y_1^* is the optimal shadow price of water and x^* is the aggregate output of crops, then the response of crop substitution due to varying water rates may be directly estimated. For public sector pricing policies, these optimal crop substitution rates are of great value in assessing the welfare implications.

Game Theory and LP Models

The most important application of saddle point inequality is in terms of two-person zero-sum games. There are two simple ways by which an LP model can be converted to a zero-sum game formulation.

Consider the LP problem given in (3.8) before and assume without loss of generality that all c_j's are positive; if they are not, a large positive number can be identically added to each c_j without altering the optimal feasible solution of the problem. Then define the following

$$p_j = c_j x_j / z; \quad \bar{a}_{ij} = a_{ij}/(b_i c_j), \quad \begin{array}{l} i=1,2,\ldots,m \\ j=1,2,\ldots,n \end{array}$$

$$z = \sum_{j=1}^{n} c_j x_j.$$

The LP model may then be reduced to

$$\max z = z \sum_{j=1}^{n} p_j$$

$$\text{s.t. } z \sum_{j=1}^{n} \bar{a}_{ij} \cdot p_j \le 1, \quad \sum_{j=1}^{n} p_j = 1, \quad p_j \ge 0; \quad \begin{array}{l} j=1,2,\ldots,n \\ i=1,2,\ldots,m. \end{array}$$

Defining $u = 1/z$ and using $\sum_{j=1}^{n} p_j = 1$ in the objective function and noting that maximizing z is equivalent to minimizing the scalar quantity u, we obtain:

Min u

$$\text{s.t. } \sum_{j=1}^{n} \bar{a}_{ij} p_j \le u; \quad \sum_{j=1}^{n} p_j = 1, \quad p_j \ge 0. \tag{4.1}$$

Likewise for the dual problem (3.11) we define

$$q_i = b_i y_i / C; \quad C = \sum_{i=1}^{m} b_i y_i, \quad \sum_{i=1}^{m} q_i = 1, \quad v = 1/C \text{ and obtain}$$

Max v

$$\text{s.t. } \sum_{i=1}^{m} q_i \bar{a}_{ij} \ge v; \quad \sum_{i=1}^{m} q_i = 1, \quad q_i \ge 0 \tag{4.2}$$

$$i=1,2,\ldots,m.$$

Note that the n-tuple vector p and m-tuple vector q can be interpreted as probability vectors and hence as mixed strategies. If $p_j = 1$ for only one j out of n possible strategies, so that $p_j = 0$ for all others then we have a pure strategy. Likewise for the model (4.2) the optimal solution vector q* is called a pure strategy if only one of m q_i's is one and the others are zero. Two aspects of the game formulation are to be noted. One is that if there is an optimal solution pair (x*,y*) of the primal and dual LP models (3.8) and (3.11), then there must exist the pair (p*,q*) of the optimal strategy vectors for (4.1) and (4.2) respectively. The reverse is also true i.e. if there exists the pair (p*,q*) solving (4.1) and (4.2) respectively, then one can construct a pair of primal and dual LP problems with optimal solutions (x*,y*). However since $\bar{a}_{ij}$ is a nonlinear combination of a_{ij}, b_i, c_j it may not be possible to recover all the three sets of parameters from one set of $\bar{a}_{ij}$ coefficients. Secondly, one notes from the first set of constraints of (4.1) that u can be viewed as

$$u = \max_{1 \leq i \leq m} (\sum_{j=1}^{n} \bar{a}_{ij} p_j)$$

Hence minimize u means

$$\min u = \min \max_{1 \leq i \leq m} (\sum_{j=1}^{n} \bar{a}_{ij} p_j)$$

In this sense p* is a minimax solution. Likewise q* is a maximin solution. Further u* = min u equals v* = max v by the strong duality of LP models. Suppose that player 1 uses his optimal strategy vector p* and player 2 the streategy vector q* in terms of their own respective LP models (4.1) and (4.2). Then player 2 maximizes his lowest payoff, lowest in the sense of any strategy chosen by his opponent player 1. In this sense the maximum solution (q*,v*) is the best of the worst. Likewise player 1 whose payoff is opposite to that of player 2 is interested in paying as little as possible to the other player; so he considers first the worst case that might happen i.e. the maximum value u that he may have to pay when the other player has hurt him most. His p* minimizes the worst case payoff. But since min u = -max u we get the sum of the two payoffs, both in the same direction

$$-u^* + v^* = -v^* + v^* = 0, \text{ since } u^* = v^*.$$

Consider the LP example in (2.1) before:

$$\max z = 2.5x_1 + 2x_2 \quad (4.3)$$
$$\text{s.t. } x_1 + 2x_2 \leq 80, \; 3x_1 + 2x_2 \leq 90$$
$$x_1 \geq 0, \; x_2 \geq 0.$$

Its dual is

$$\min C = 80y_1 + 90y_2$$
$$\text{s.t.} \quad (4.4)$$
$$y_1 + 3y_2 \geq 2.5, \; 2y_1 + 2y_2 \geq 2$$
$$y_1 \geq 0, \; y_2 \geq 0.$$

Hence the two equivalent game formulations due to (4.1), (4.2) are as follows:

$$\min u$$
$$\text{s.t.} \quad 0.005\, p_1 + 0.012\, p_2 \leq u$$
$$0.013\, p_1 + 0.011\, p_2 \leq u \quad (4.5)$$
$$p_1 + p_2 = 1, \quad p_1, p_2 \geq 0$$

and max v
s.t. $\quad 0.005\ q_1 + 0.013\ q_2 \geq v$
$\quad\quad\ 0.012\ q_1 + 0.011\ q_2 \geq v$ (4.6)
$\quad\quad\ q_1 + q_2 = 1, \quad q_1, q_2 \geq 0.$

We know that both $x_1^* = 5$, $x_2^* = 5$ are positive, hence p_1^*, p_2^* are proper fractions i.e. mixed strategies, so are the q_1^* and q_2^*. Hence we can equate the first two inequalities in both (4.5) and (4.6) and obtain

$$0.005\ p_1^* + 0.012\ p_2^* = u^* = 0.013\ p_1^* + 0.011\ p_2^*$$
$$\text{or,}\quad p_1^* = 1/9,\ p_2^* = 8/9;\ u^* = 0.011$$

and

$$0.005\ q_1^* + 0.013\ q_2 = v^* = 0.012\ q_1 + 0.011\ q_2^*$$
$$\text{or,}\quad q_1^* = 2/9,\ q_2^* = 7/9;\ v^* = 0.011.$$

A second method of showing the equivalence of a zero-sum game to an appropriate LP model is to consider an arbitrary two-person game in a normal form and show that under certain conditions it leads to an LP formulation.

Let b_{ij} be the payoff to player 1 when he chooses his strategy p_j and player 1 chooses q_i. Then the expected payoff to player 1 is

$$\sum_{j=1}^{n}\left(\sum_{i=1}^{m} q_i\ b_{ij}\right) p_j = \sum_{j=1}^{n} r_j\ p_j,\ r_j = \sum_{i=1}^{m} q_i\ b_{ij},\text{ if there are m possible}$$

strategy choices for player 2 and n for player 1. Thus player 1 has to optimally choose his strategy vector p by maximizing his expected payoff:

$$\underset{p}{\text{Max}}\ J(p) = \sum_{j=1}^{n}\sum_{i=1}^{m} q_i\ b_{ij}\ p_j$$

$$\text{s.t.}\quad \sum_{j=1}^{n} p_j = 1,\quad p_j \geq 0;\quad j=1,2,\ldots,n.$$ (4.7)

Similarly, if c_{ij} is the payoff to player 2 when 1 selects p_j and 2 selects q_i, then the optimal decision for player 2 is to maximize his own expected payoff

$$J(q) = \sum_{i=1}^{m} q_i\left(\sum_{j=1}^{n} c_{ij}\ p_j\right)$$

by choosing his own strategies $q_1,\ldots,q_m$. This leads to the model:

$$\underset{q}{\text{Max}}\ J(q) = \sum_{i=1}^{m}\sum_{j=1}^{n} q_i\ c_{ij}\ p_j$$

$$\text{s.t.}\quad \sum_{i=1}^{m} q_i = 1,\quad q_i \geq 0;\quad i=1,2,\ldots,m.$$ (4.8)

Now the two-person game is zero-sum if $c_{ij} = -b_{ij}$ for all i and j. But if $b_{ij} + c_{ij} = k$, a constant for all i and j then it is a constant sum game. In all other cases the game is non-zero sum. In case of zero-sum game the decision model of player 1 may clearly be written as

$$\text{Min } J(q) = \sum_{i=1}^{m} \sum_{j=1}^{n} q_i b_{ij} p_j \tag{4.9}$$

$$\text{s.t.} \quad \sum_{i=1}^{m} q_i = 1, \quad q_i \geq 0, \quad i=1,2,\ldots,m.$$

Assume that player 1 announces his optimal strategy vector $p^* = (p_j^*)$. Given this knowledge, the best that player 2 can do in (4.9) is to minimize his loss

$$J(q) = \sum_{i=1}^{m} q_i w_i, \text{ where } w_i = \sum_{j=1}^{n} b_{ij} p_j^*.$$

Hence he chooses his optimal best reply $q^* = (q_i^*)$. The sum of the two expected payoffs is $\sum_{i=1}^{m} \sum_{j=1}^{n} q_i^*(b_{ij} + c_{ij}) p_j^* = 0$,

since $c_{ij} = -b_{ij}$.

Clearly this formulation is more general, since it can incorporate cooperative game situations where one's payoff is not at the expense of his rivals i.e. the game is positive sum. By this standard a zero-sum two-person game is one of extreme conflict.

2.2 Nonlinear Programming

Two types of models in nonlinear programming (NLP) are useful in applied economics. One is a quadratic program (QP) where the objective function is quadratic but the restrictions are linear. The second is a concave (convex) programming model, where the restrictions are still linear but the objective function is a concave (convex) nonlinear function of the decision vector x for the maximization (minimization) problem. Both these cases are covered by the Kuhn-Tucker (KT) Theorem, which gives a set of necessary and sufficient conditions for the optimal solution. But it is important to note that there exists other cases of NLP not covered by the KT Theorem.

2.2.1 Kuhn-Tucker Theory

The Kuhn-Tucker theory in its simplest version of NLP considers maximizing a concave nonlinear function $f(x) = f(x_1,\ldots,x_n)$ of n variables x_j subject to a constraint set R defined by a set of linear inequalities and nonnegativity on x i.e.

$$\underset{x \in R}{\text{Max }} f(x), \text{ where } R = \{x \mid Ax \leq b, x \geq 0\} \tag{5.1}$$

In case of minimization the objective function $f(x)$ must be convex. If the constraint set R also involves concave nonlinear functions $h_i(x)$ such that

$$R = \{x \mid x \geq 0;\ h_i(x) = b_i - g_i(x_1, x_2, \ldots, x_n) \geq 0 \qquad i=1,2,\ldots,m\} \tag{5.2}$$

then we have a general case of NLP analyzed by the KT theory.

Quadratic Program (QP)

We first consider an example of the simplest version (5.1) which is known as a quadratic programming (QP) model, which occurs quite frequently in economics. For instance the objective function of an LP model when viewed as profits for the multi-product firm $z = c^T x = (p - v)^T x$ with $c_j = p_j - v_j$ where p_j = the price per unit of output x_j, v_j = all variable costs per unit of output x_j, will be reduced to a quadratic form, if the selling market faces imperfect competition. Assuming the n markets to be independent so that the demand functions are separable as

$$p_j = g_j - h_j x_j, \qquad h_j > 0, \qquad j=1,2,\ldots,n \tag{5.3}$$

then $c_j = (g_j - v_j) - h_j x_j$ and the profit function becomes

$$z = \sum_{j=1}^{n} [(g_j - v_j) - h_j x_j] x_j \tag{5.4}$$

which is quadratic and <u>strictly concave</u> since the second derivative of z with respect to x_j is <u>negative</u>. In case the demand functions are not separable, so that demand in market j is influenced by prices in market k, $k \neq j$, then the inverse demand function (5.3) appears as

$$p_j = g_j - \sum_{k=1}^{n} h_{jk} x_k$$

and the objective function becomes

$$z = \sum_{j=1}^{n} (g_j - v_j) x_j - \sum_{j=1}^{n} \sum_{k=1}^{n} x_j h_{jk} x_k \tag{5.5}$$

In this case the objective function is

(i) strictly (weakly) concave, if the matrix $H = (-h_{jk})$ is negative definite (negative semi-definite).
(ii) strictly (weakly) convex, if the matrix H is positive definite (positive semidefinite), or
(iii) neither concave or convex if the matrix H is not definite at all (i.e. neither negative semidefinite, nor positive semidefinite).

If the restriction set R defined in (5.1) is not empty, then the KT theory gives the necessary and sufficient conditions for a maximizing solution x^* of (5.5) subject to $x \in R$. Note that cases (ii) and (iii) are not covered by the KT theory-i.e. the maximizing solution in these two cases may not satisfy the KT necessary conditions at all.

A second economic example of QP arises in Markowitz-Tobin portfolio theory. In its simple version a portfolio model considers an investor who has to optimally invest a proportion x_i of his total wealth in a risky asset i for i=1,2,...,n where $x_i \geq 0$, and $\sum_{i=1}^{n} x_i = 1$. The return r_i on each risky asset i traded in the competitive stock market is random i.e. subject to probabilistic fluctuations in the daily stock market. Its expected or mean return $m_i = E(r_i)$ may be calculated over a fixed period and also its variance and covariances as $v_{ij} = E[(r_i - m_i)(r_j - m_j)]$ where E denotes expectation. If the different assets are mutually independent i.e. their covariances of returns are zero, then any portfolio-mix of n securities $x^P = (x_1^P, x_2^P, \ldots, x_n^P)$ would have an expected reutrn $R_p = \sum_{i=1}^{n} m_i x_i^P$ and total variance $V_p = \sum_{i=1}^{n} x_i^P v_{ii} x_i^P$, since the covariances v_{ij} are all zero for $i \neq j$. Since the fluctuations of returns in a portfolio are measured by the overall variance V_p, an optimal or efficient portfolio is defined in Markowitz theory by minimizing the overall variance V_p subject to a lower bound on expected return i.e. $R_p \geq c$, where c is a positive constant preassigned by the investor's subjective judgement. In the general case when the covariances are not zero we have the QP model for determining the optimal allocations $x_1, x_2, \ldots, x_n$ as follows:

$$\min_{x \in R} V_p = \sum_{i=1}^{n} \sum_{j=1}^{n} x_i v_{ij} x_j$$

where

$$R = \{x \mid \sum_{i=1}^{n} m_i x_i \geq c; \quad \sum_{i=1}^{n} x_i = 1; \quad x_i \geq 0; i=1,\ldots,n\}$$

Since by statistical conditions the variance-covariance matrix $V = (v_{ij})$ must be positive semidefinite, only two cases are possible, both of which are covered by the KT theory.

(i) the covariance matrix V is positive definite, then the objective function is strictly convex and hence the minimizing vector $x^* = (x_i^*)$ is unique, and

(ii) the covariance matrix is positive semidefinite, in which case the objective function is only convex (and not strictly convex) and hence the minimizing vector x^* may not be unique.

In case of the NLP (5.1) where the objective function $f(x)$ is concave and differentiable, let us assume that the constraint set R is nonempty. Let $x^* = (x_j^*)$ be the optimal solution vector. Then the KT Theorem states the following.

Kuhn-Tucker Theorem

There exists an m-tuple vector $y^* = (y_i^*)$ of nonnegative Lagrange multipliers associated with each inequality constraint

$$\sum_{j=1}^{n} a_{ij} x_j \leq b_i; \qquad i=1,2,\ldots,m$$

such that the pair (x^*, y^*) of optimal vectors satisfy the following two necessary conditions

(1) $(\partial L/\partial x_j)^* \leq 0$, $(x_j \partial L/\partial x_j)^* = 0$, $x_j^* \geq 0 \qquad j=1,2,\ldots,n$

(2) $(\partial L/\partial y_i)^* \geq 0$, $(y_i \partial L/\partial y_i)^* = 0$, $y_i^* \geq 0 \qquad i=1,2,\ldots,m$

where $L = L(x,y) = f(x_1, x_2, \ldots, x_n) + \sum_{i=1}^{m} y_i (b_i - \sum_{j=1}^{n} a_{ij} x_j)$ is the scalar Lagrangean function of (n+m) variables x_j's and y_i's.

The two sufficiency conditions, which are necessarily fulfilled in this case are the following:

(3) $L(x, y^*)$ is concave in x around the optimal point x^*, when y is fixed at y^*, and

(4) $L(x^*, y)$ is convex in y around the optimal point y^*, when x is fixed at x^*.

Interpretations

Some economic interpretations of the KT Theorem would be most helpful at this stage. First of all, the Lagrangean $L(\cdot)$ viewed as a function of x is net profits and it is net total costs when viewed as a function of y. Hence the necessary condition (1) implies that if we deviate away from the optimal point x_j^* we have extra profits negative or zero. Extra profit at the optimal point is negative means that any movement in a small neighborhood of x^* yields less total profits and i.e.

it is unprofitable. Extra profit is zero implies that there exist other points nearby yielding identical profits as at x^* i.e. there exist more than one optima giving an identical level of maximum profits. Similarly, the necessary condition (2) implies that any change in y_i in a small neighborhood of the optimal point y^*, which is a minimizer of the Lagrangean function $L(x^*,y)$, would not lead to any decrease in total cost from the minimal level $L(x^*,y^*)$. It is clear that for nonnegative vectors x, y one can give a saddle point characterization as follows:

$$L(x,y^*) \leq L(x^*,y^*) \leq L(x^*,y).$$

Secondly, the second part of the two necessary conditions has the economic interpretation that any activity (or variables) which yields marginal loss should not be used at all. Thus

if $(\partial L/\partial x_j)^* < 0$, then $x_j^* = 0$, $j=1,2,\ldots,n$

and

if $(\partial L/\partial y_i)^* > 0$, then $y_i^* = 0$; $i=1,2,\ldots,n$.

As in LP, the second is the shadow pricing rule i.e.

$$(\partial L/\partial y_i)^* = b_i - \sum_{j=1}^{n} a_{ij} x_j^* > 0$$

implies a positive surplus at the optimal point, hence its optimal shadow price (y_i^*) must be zero. The first part in

$$(\frac{\partial L}{\partial x_j})^* = (\frac{\partial f(x)}{\partial x_j})^* - \sum_{i=1}^{m} y_i^* a_{ij} < 0$$

implies that total marginal cost (MC) at the optimal point $MC(y^*) = \sum_{i=1}^{m} y_i^* a_{ij}$ exceeds marginal revenue $MR(x^*) = (\partial f/\partial x_j)^*$, hence this leads to marginal loss i.e. x_j^* should not be produced at all.

Thirdly, the two sufficiency conditions (3), (4) imply that there exist two neighborhoods around the optimal points x^*, y^* respectively, where the Lagrangean can be iterated to find the saddle point i.e. there exists a maximizing direction from an x in this neighborhood ($x \rightarrow x^*$) and a minimizing direction from a y in its own neighborhood (i.e. $y \rightarrow y^*$). Note that the optimal shadow prices y_i^* can be used in various policy formulations exactly as in linear programming i.e. y_i^* can be interpreted as optimal marginal profitability of a resource b_i i.e. $y_i^* = \partial f(x^*)/\partial b_i$. Hence if the resource b_i is not used to the full at the optimal profit level, then y_i^* is zero.

A QP model is one popular form of concave programming with linear restrictions because with very slight modifications the simplex algorithms of LP models do indeed apply. Computer routines are therefore easily available especially for the case of a strictly concave quadratic objective function.

In the more general case of NLP problems, the restriction set R defined by (5.2) also involves differentiable concave functions $h_i(x) = b_i - g_i(x)$, $i=1,2,\ldots,m$ along with the concave objective function $f(x)$ to be maximized. Here the KT theory requires that a special condition known as <u>constraint qualifications</u> (CQ) be satisfied, before the necessary and sufficient conditions of the KT Theorem are applied to characterize an optimal point (x^*, y^*).

Constraint Qualifications (CQ)

There are two simple intuitive ways to understand the meaning of the CQ test. The first is that it tests the nonsingularity of the boundary points defined by those constraints that are equalities i.e. the effective constraints. The second is that this nonsingularity may be tested by a simple <u>rank condition test</u> applied to the effective constraints only. For example consider the problem

$$\max \ f(x_1, x_2) = x_1$$
$$\text{s.t.} \ h_1(x_1, x_2) = (1 - x_1)^3 - x_2 \geq 0$$
$$x_1 \geq 0, \ x_2 \geq 0.$$

Since the objective function is to maximize x_1 where $x_2 \leq (1 - x_1)^3$, clearly the optimal solution is at the point $x_1^o = 1$, $x_2^o = 0$. Two ways we can test if CQ is satisfied at this point $(1,0)$. First, we apply the KT necessary condition on the Lagrangean function $L = x_1 + y_1 ((1 - x_1)^3 - x_2)$:

$$\frac{\partial L}{\partial x_1} = 1 - 3 y_1 (1 - x_1)^2 \leq 0. \tag{5.6}$$

But since $x_1^o = 1$ as above gives a positive value of x_1, the complementary slackness rule of KT theory gives $\partial L / \partial x_1 = 0$ at $x_1 = x_1^o = 1$. But by (5.6) $\partial L / \partial x_1 = 1$ which cannot be equal to zero. The reason for this inconsistency is due to the singularity at the boundary point $(1,0)$. A second way to test the satisfaction of CQ is to note that at the point $(1,0)$ we have only two <u>effective</u> constraints $(1 - x_1^o)^3 - x_2^o = 0$, $x_2^o = 0$ which can be combined to one: $h_1(x_1^o, x_2^o) = (1 - x_1^o)^3 = 0$. To test by the rank condition we take its derivative

$$\frac{\partial h_1}{\partial x_1} = -3(1 - x_1^o)$$

which equals zero at (1,0) i.e. the rank of $\partial h_1/\partial x_1$ is zero. Hence the point (1.0) is singular i.e. it defines an outward pointing <u>cusp</u> not belonging to the feasible region, which can invalidate the KT Theorem. Take a second example

$$\max f = x_1 x_2$$
$$\text{s.t.} \quad h_1 = \tfrac{1}{2}(x_1^2 + x_2^2) - 1 = 0$$

By applying the first order condition to the Lagrangean function $L = x_1 x_2 + y_1(\tfrac{1}{2}(x_1^2 + x_2^2) - 1)$ we get

$$\frac{\partial L}{\partial x_1} = x_2 + y_1 x_1 = 0, \quad \frac{\partial L}{\partial x_2} = x_1 + y_1 x_2 = 0, \quad \frac{\partial L}{\partial y_1} = \tfrac{1}{2}(x_1^2 + x_2^2) - 1 = 0.$$

Clearly, the values $x_1^o = 1$, $x_2^o = -1$, $y_1^o = 1$ satisfy the first order condition. To apply the rank condition rule at the point $x^o = (1,-1)$ we compute the two partial derivatives

$$\frac{\partial h_1}{\partial x_1} = x_1^o = 1, \quad \frac{\partial h_1}{\partial x_2} = x_2^o = -1$$

and check if the rank of $(\frac{\partial h_1}{\partial x_1}, \frac{\partial h_1}{\partial x_2}) = (1, -1)$ is equal to one, since we have only one effective constraint. Obviously the rank is one and hence the CQ holds at (1,-1). In the general case we can therefore say that if there are m effective constraints $h_i(x) = 0$, $i=1,2,\ldots,m$ and n variables x_j ($j=1,2,\ldots,n$), with m < n we form the n by n matrix $R = \partial h_i(x)/\partial x_j$ and evaluate it at a vector point x^o. If the rank of this matrix R at x^o equals m, the number of effective constraints then the CQ is satisfied. If the rank of R is less than m, then the CQ fails. In case the CQ fails, then the KT necessary conditions may not necessarily characterize the optimal solution.

For most practical cases one could ask if all the constraints $h_i(x) \geq 0$, $i=1,2,\ldots,m$ are concave and differentiable at any vector point $x \geq 0$, where at least one component is positive. If they are, then the CQ is satisfied. Why? Because one may start from such a boundary point through an arc inwardly to the feasible set; also by concavity the rank condition rule must be satisfied for every non-trivial boundary point.

<u>General Version of KT Theorem</u>

One has to note that for maximization (minimization) problems, the KT Theorem requires that the objective function be concave (convex) and differentiable. Hence the theorem does not apply if we have to

$$\text{maximize} \quad f(x) = (x - 1)^3$$
$$\text{s.t.} \quad 2 - x \geq 0, \quad x \geq 0$$

where x is a scalar. The values $x° = 1$, $\lambda° = 0$ where $\lambda°$ is the optimal value of the Lagrange multiplier satisfy the KT necessary conditions since

$$\frac{\partial L}{\partial x} = 3(x-1)^2 - \lambda \leq 0 \quad \text{for } x° = 1, \lambda° = 0$$

$$\frac{\partial L}{\partial \lambda} = 2-x \geq 0 \quad \text{at } x° = 1, \lambda° = 0$$

but the true optimal values are given by $x° = 2$, $\lambda° = 0$. Why did the KT necessary conditions fail? The objective function is convex for all $x > 1$.

Another example is

$$\max f(x_1, x_2) = \tfrac{1}{2}(x_1^2 + x_2^2) \quad \text{s.t.} \quad x_1 + x_2 = 1$$
$$x_1 \geq 0, x_2 \geq 0$$

Since the objective function is strictly convex in x_1, x_2 it has a minimum, but the objective is to maximize and not to minimize. The KT necessary conditions fail to identify the maximum which is given by $x_1 = 1$, $x_2 = 0$ or, by $x_1 = 0$, $x_2 = 1$ whereas the minimizing values are $x_1 = x_2 = 1/2$.

For the NLP problem

$$\underset{x}{\text{Max}}\ f(x) = f(x_1, x_2, \ldots, x_n) \tag{6.1}$$

$$\text{s.t.} \quad h_1(x) = b_1 - g_1(x_1, \ldots, x_n) \geq 0$$
$$h_2(x) = b_2 - g_2(x_1, \ldots, x_n) \geq 0 \tag{6.2}$$
$$\vdots$$
$$h_m(x) = b_m - g_m(x_1, \ldots, x_n) \geq 0$$
$$x_1 \geq 0, x_2 \geq 0, \ldots, x_n \geq 0$$

the most general form of the KT Theorem may be stated as follows.

Generalized KT Theorem

In the NLP problem (6.1), (6.2) with functions $f(x), h_1(x), h_2(x), \ldots, h_m(x)$ differentiable, let the objective function $f(x)$ be <u>pseudoconcave</u> and each of the m constraints $h_i(x)$ be <u>quasi-concave</u>. Suppose the KT necessary conditions hold for a vector x^*. Then this vector point x^* is also optimal for the NLP problem.

To understand this generalized version we need to define <u>pseudoconcavity</u> and <u>quasi-concavity</u>. A differentiable scalar function $f(x)$ is pseudoconcave or one-sided concave if the function continues to decrease in that direction whenever a directional derivative indicates a decrease. Thus $f(x_1,x_2)$ is pseudoconcave in two variables x_1, x_2 around a specified point (x_1^o, x_2^o) if

$$\left(\frac{\partial f(x)}{\partial x_1}\right)_{x^o} (x_1 - x_1^o) + \left(\frac{\partial f(x)}{\partial x_2}\right)_{x^o} (x_2 - x_2^o) \leq 0$$

implies

$$f(x_1,x_2) \leq f(x_1^o,x_2^o)$$

A function $f(x_1,x_2)$ of two variables is defined to be <u>quasi-concave</u> in a certain domain D, if for any scalar c, the set

$$H_c = \{x_1, x_2 \mid f(x_1, x_2) \geq c\} \tag{6.3}$$

is convex. A set of points is convex, if every chord joining any two points of the set lies entirely in the set.

Some authors do not distinguish between pseudoconcavity and quasi-concavity and state the generalized KT Theorem in terms of the quasi-concave objective function $f(x)$ and the quasi-concave constraints $h_i(x) \geq 0$, $i=1,2,\ldots,m$ defined in (6.1) and (6.2) before.

2.2.2 Economic Applications

All the economic applications we discussed before in connection with LP problems e.g. (a) peak-load pricing, (b) use of shadow prices in competitive equilibrium and (c) saddle point inequality of game theory apply in case of NLP models. Two additional applications may be mentioned here, one in optimal portfolio theory and the other in optimal production decisions under price uncertainty.

<u>Ex 6</u> (optimal portfolio selection model)

An investor has to allocate his wealth of $1.0 say to buy two securities having returns r_1 and r_2. Let x_1, x_2 be the nonnegative proportions allocated to the two securities where $x_1 + x_2 = 1$. Any choice of x_1, x_2 with the restrictions $x_1 \geq 0$, $x_2 \geq 0$ and $x_1 + x_2 = 1$ is called a portfolio. The return r_i for each security i is random in the sense that its future value is unknown. Based on the past realized values its mean m_i and variances v_i and covariances v_{ij} may be computed as follows

$$m_i = \sum_{t=1}^{T} r_{it}/T, \quad v_i = \sum_{t=1}^{T} (r_{it} - m_i)^2/T$$

$$v_{ij} = \sum_{t=1}^{T} [(r_{it} - m_i)(r_{jt} - m_j)]/T$$

where T is the number of past observations and r_{it} is the realized return of security i at time t. The mean-variance model seeks to determine the optimal portfolio by solving the following quadratic programming problem

$$\underset{x_1, x_2}{\text{Minimize}} \quad \sigma_p^2 = \sum_{i=1}^{2} \sum_{j=1}^{2} x_i v_{ij} x_j$$

$$\text{s.t.} \quad \sum_{i=1}^{2} m_i x_i \geq c$$

$$\sum_{i=1}^{2} x_i = 1, \qquad x_i \geq 0, \ i=1,2$$

where σ_p^2 is the total variance of returns on the portfolio, which measures the total risk due to fluctuations of returns around their mean values and $R_p = \sum_{i=1}^{2} m_i x_i$ denotes the total expected return on the portfolio. Here c denotes a positive number representing minimal expected return below which the investor may not be willing to take the risk of investing in the stock market. The above model is readily extended to n securities by letting i and j extend from 1 to n i.e. in vector-matrix notation

$$\underset{x}{\text{Min}} \ \sigma_p^2 = x^T V x$$

$$\text{s.t.} \quad m^T x \geq c$$
$$e^T x = 1, \quad x \geq 0$$

where x, m, e are n-tuple column vectors with typical elements x_i, m_i and e_i, where each element e_i of vector e is one and $V = (v_{ij})$ is a square matrix of order n denoting the covariances of all pairs of n securities.

For any fixed positive value of c, this NLP defines a convex programming problem which can be written as a concave program by maximizing $(-\sigma_p^2) = -x^T V x$. By statistical properties the variance-covariance matrix V is positive semi-definite (and usually positive definite), hence σ_p^2 is convex (strictly convex) and $(-\sigma_p^2)$ is concave (strictly concave) in vector x. Also the restrictions of the model are all linear and hence the Kuhn-Tucker constraint qualifications are necessarily satisfied. The KT Theorem may therefore be used to characterize and compute the optimal solution vector x*.

To get an explicit idea of the optimal solution, one may consider a simpler model by ignoring the nonnegativity condition on x and writing the minimal expected return condition as an equality i.e.

$$\text{Min } \sigma_p^2 = x^T V x$$
$$\text{s.t. } m^T x = c, \quad e^T x = 1$$

By applying the usual necessary condition of calculus on the Lagrangean function

$$L(x, y_1, y_2) = x^T V x + y_1(m^T x - c) + y_2(e^T x - 1)$$

we set
$$\partial L/\partial x_j = 0, \quad j=1,2,\ldots,n$$
$$\partial L/\partial y_1 = 0, \quad \partial L/\partial y_2 = 0.$$

Assuming V to be positive definite, so that its inverse V^{-1} exists, the optimal solution vector x^* and the associated objective function σ_p^{*2} may be computed as follows

$$x^* = \frac{V^{-1}}{(ad-b^2)} [(md - b)c + (ae - bm)]$$

$$\sigma_p^{*2} = \min \sigma_p^2 = \frac{dc^2 - 2bc + a}{ad - b^2}$$

where $a = m^T V^{-1} m$, $b = m^T V^{-1} e$ and $d = e^T V^{-1} e$. By varying c on the positive axis one can generate the whole portfolio efficiency frontier for a fixed parameter set (m,V). Note that the minimal variance σ_p^{*2} is a strictly convex function of c since d is always positive. This implies that σ_p^{*2} can be further minimized by suitably choosing the parameter c whenever it is realistic to do so for some risk-averse investors.

This type of portfolio model can be generalized to characterize an efficient capital market behavior.

Ex 7 (optimal production under price uncertainty)

A farmer produces two crops x_1, x_2 with net returns c_1, c_2 per unit so that his total net return is $z = c_1 x_1 + c_2 x_2$ but since future prices of crops are unknown (i.e. random in this case) he can only use the mean values $\bar{c}_1$, $\bar{c}_2$ of net returns from his past observations. If the returns of the two crops are statistically independent i.e. not correlated, then their covariances are zero; hence the variance of total net return would be given by $\sigma_z^2 = v_{11} x_1^2 + v_{22} x_2^2$ where v_{ii} is the variance of return of crop i. The total mean return is $\bar{z} = \bar{c}_1 x_1 + \bar{c}_2 x_2$. Assume that there are two resource constraints due to land (b_1) and capital (b_2) i.e.

$$a_{11}x_1 + a_{12}x_2 \leq b_1 \text{ (land availability)}$$
$$a_{21}x_1 + a_{22}x_2 \leq b_2 \text{ (capital limit)} \quad (7.1)$$
$$x_1 \geq 0, \; x_2 \geq 0 \quad \text{(nonnegativity of output)}$$

where a_{ij}'s are the input-output coefficients. What would be the optimal crop mix x_1^* and x_2^*? Several possibilities exist. The farmer may ignore the variance term σ_z^2 altogehter; he is then said to ignore the risk associated with the random fluctuation of c_i values around their means $\bar{c}_i$. His decision model is then

$$\text{Max } \bar{z} = \bar{c}_1 x_1 + \bar{c}_2 x_2$$
$$\text{subject to (7.1).}$$

This is a standard LP model. Alternatively, the farmer may be risk-averse in the sense that he is aware that the actual profit may fall below the mean level $\bar{z}$ in many cases and hence he is willing to maximize a risk-adjusted profit level z_A given by $z_A = \bar{z} - \alpha \sigma_z^2$ where α is a positive number. The higher the value of α, the more risk-averse the farmer is. Two interpretations of the term $\alpha \sigma_z^2$ are available. One is that it denotes the cost of risk i.e. if due to high variability, net returns fell much below the average level $\bar{z}$, the farmer may go broke. A second view is that by choosing the goal of adjusted return z_A lower than that of the mean level $\bar{z}$, the farmer may substantially increase the probability of attaining his goal i.e. it defines a safety first policy. For the case of n crops this leads to a quadratic programming model as follows, which has been very widely applied in agricultural studies:

$$\text{Max } z_A = \sum_{j=1}^{n} \bar{c}_j x_j - \alpha \sum_{i=1}^{n} \sum_{j=1}^{n} x_i v_{ij} x_j$$

$$\text{s.t.} \quad \sum_{j=1}^{n} a_{ij} x_j \leq b_i; \quad i=1,2,\ldots,m \quad (7.2)$$

$$x_j \geq 0, \; j=1,2,\ldots,n.$$

Other approaches leading to more general NLP models, which utilize the probability distribution of the profit function z are discussed in the theory of stochastic linear programming.

2.2.3 Set-theoretic Concepts of Efficiency

The concept of an optimum may be characterized in a very abstract way in terms of set theory. Although this theory is abstract, it has two important operational implications. One is that it can characterize some situations where the optimum solution exists (i.e. existence) and if so, it can derive its duality or shadow price implications. Hence it is widely used to characterize a general equilibrium competitive model and Pareto optimality. Hence we consider a brief introduction to sets.

A set is a collection of objects. Thus if S is a set of all positive integers we denote it by S = {x: x is a positive integer}. If n is a member of a set S we denote it by n ∈ S. If n is not a member of S, then we denote it by n ∉ S. A set can be specified in many dimensions. Thus the set S of points on the unit circle in the two-dimensional plane can be denoted by S = {(x,y): $x^2 + y^2 = 1$ for x ∈ R, y ∈ R} where R is the set of all real numbers. A set S may be discrete i.e. S = {1,3,5} or continuous i.e S = {x: $1 \leq x \leq 6$}. Given two sets A and B, if every elemnt of A is an element of B, we say that A is a subset of B. Thus two sets A and B are equal if they have the same elements. The following types of sets are most often used in applied mathematics.

(i) <u>Intersection</u>: the collection of all the elements which belong to both set A and set B is called the <u>intersection</u> and denoted by A ∩ B. This A ∩ B = {x: x ∈ A and x ∈ B}. For example, if A = {x: $1 \leq x \leq 6$} and B = {x: $3 \leq x \leq 7$} then A ∩ B = {x: $3 \leq x \leq 6$}.

(ii) <u>Union</u>: the collection of all the elements which belong to <u>either</u> set A <u>or</u> set B is called the union of the sets A and B and denoted by A ∪ B. Thus, if A = {x: $1 \leq x \leq 6$} and B = {x: $3 \leq x \leq 7$}, then A ∪ B = {x: $1 \leq x \leq 7$}.

(iii) <u>Null set</u>: the set that has no elements is called an empty or <u>null</u> set and denoted by ϕ.

(iv) <u>Disjoint</u>: if two sets have no points in common, they are called <u>disjoint</u> or nonintersecting, i.e. A ∩ B = ϕ. Thus A = {x: $1 \leq x \leq 6$} and B = {x: $7 \leq x \leq 15$} are disjoint sets.

(v) <u>Neighborhood</u>: an <u>epsilon</u> neighborhood of a vector point p = (p_1, p_2) in two dimensions is specified by the set N (p, ε), where ε is a positive scalar

$$N(p, \varepsilon) = \{x: \sqrt{\{(x_1 - p_1)^2 + (x_2 - p_2)^2\}} < \varepsilon\}.$$

Since the Euclidean distance of point x from point p is defined in n dimensions by

$$d(x,p) = \sqrt{\sum_{i=1}^{n} (x_i - p_i)^2}$$

the epsilon neighborhood of p in n dimensions can be written as

$$N(p, \varepsilon) = \{x: d(x,p) < \varepsilon\}.$$

In two dimensions if $p = (0,0)$ is the origin and $\varepsilon = 1$, then the set

$$N = \{(x_1, x_2): \sqrt{(x_1^2 + x_2^2)} < 1\}$$

specifies a disc of radius one with the center at the origin. Note that this set N has <u>boundary</u> points on the circle of radius 1, but the boundary is not included in the set. This means the set N is open. Mathematically, x is a boundary point of a set S if every neighborhood of x contains at least one point in S and one point not in S. Here 'every' means that the positive scalar ε may be as small as possible but staying above zero.

(vi) <u>Closed set</u>: a set is closed if it contains all its boundary points. Thus the set $S = \{x: 1 \leq x \leq 4\}$ is closed. Any finite union or intersection of closed sets is closed.

(vii) <u>Open set</u>: a set is open if it is possible to find some neighborhood of every one of its points which is interior i.e. completely contained in the set. Thus the epsilon neighborhood $N(p,\varepsilon)$ defined above is an open set.

(viii) <u>Convex set</u>: a set S of vector points of any dimension n is <u>convex</u>, if the chord joining <u>every</u> <u>pair</u> (x,y) of points belonging to the set is also contained in the set. Since the chord or line joining the two points x, y is $\alpha x + (1-\alpha)y$ for any value of the scalar α where $0 \leq \alpha \leq 1$, any point $z = \alpha x + (1-\alpha)y$ on this chord must belong to the set S, if $x \in S$ and $y \in S$ and $0 \leq \alpha \leq 1$.

Let R be the set of all real numbers and S be a proper subset of R. Then S is said to be <u>bounded</u> <u>from</u> <u>above</u> if there exists a scalar $a \in R$ (but a is not necessarily in S) such that $x \leq a$ for all points $x \in S$ and a is called the <u>upper</u> <u>bound</u> of S. Similarly, the scalar b is called a <u>lower</u> <u>bound</u> of the set S if there exists of $b \in R$ (but b is not necessarily in S) such that $x \geq b$ for all $x \in S$. The set S is here said to be bounded from below. Of all the upper (lower) bounds of the set S, the least (greatest) is called the supremum or sup S (the infinimum or inf S). If the supremum (infinimum) is contained in the set S, then it is called the maximum (minimum) element of the set S. By a bounded set we generally mean bounded both ways from below and above.

The set $S = \{(x_1, x_2): x_1 + x_2 \leq 1, 2x_1 + 3x_2 \leq 4, x_1 \geq 0, x_2 \geq 0\}$ defined by the linear restrictions of an LP model in two dimensions is both closed, bounded and convex i.e. compact. Similarly in the n-dimensional case if the m-tuple vector b has finite elements, the set $S = \{x: Ax \leq b, x \geq 0\}$ is also compact i.e. closed, convex and bounded. To prove convexity, let $x(1)$, $x(2)$ be two vector points in S i.e. $Ax(1) \leq b$, $x(1) \geq 0$ and $Ax(2) \leq b$, $x(2) \geq 0$. Define a new point x^o by the chord $x^o = \alpha\, x(1) + (1-\alpha)\, x(2)$ for any α, $0 \leq \alpha \leq 1$. Clearly $x^o \in S$ for any α. To see this

$$\alpha\, Ax(1) \leq \alpha b, \quad (1-\alpha)\, Ax(2) \leq (1-\alpha)b.$$

Adding $\alpha Ax(1) + (1-\alpha) Ax(2) \leq \alpha b + (1-\alpha) b = b$. Also, $\alpha x(1) + (1-\alpha) x(2) = x^o \geq 0$ since α is a nonnegative number.

Two of the most important results on convex sets are stated in the following two theorems

Weierstrass Theorem

This says that a continuous real valued function $f(x)$ on a nonempty closed bounded set S has a maximum and a minimum.

Minkowski Theorem

Let X and Y be two nonempty convex sets in n-dimensional vector points (not necessarily closed) such that they are mutually disjoint. Then there exists an n-tuple vector p of nonzero prices and a scalar α such that $p^T x \geq \alpha$ for all $x \in X$ and $p^T y \leq \alpha$ for all $y \in Y$.

The first theorem establishes the existence of an optimum solution in the set S without requiring concavity or convexity of the function $f(x)$ which could be the objective function. Hence this is more general than the KT Theorem, although it does not suggest any method of calculating the optimal point. This theorem is often used in game theory models to establish the existence of an equilibrium point.

The second theorem, also called separation theorem merely says that if X and Y are two non-empty convex sets such that $X \cap Y = \phi$ (null), then it is possible to draw a straight line (or plane in n dimensions) which has all of X on one side and all of T on the other. In other words, there is a separating hyperplane between the two disjoint sets.

As a corollary of the Minkowski Theorem, let us assume the set Y to be the origin and X be a nonempty closed convex set in n-dimensions not containing the origin so that $X \cap Y = \phi$. Then there exists an n-tuple vector p of nonzero prices and a scalar α, $\alpha > 0$ such that

$$p^T x \geq \alpha \text{ for all } x \in X$$

and this inequality can be made strict. To show that this theorem may be used to characterize an efficient point in a general production set, we first define Y as the set of all technically feasible production processes in a given economy. We assume Y to be a subset of R^n, the n-dimensional real space and the n-tuple vector y $\in$ Y specifies a particular process in this economy. We adopt the convention that the i-th component y_i of y is an output if $y_i > 0$ and an input if $y_i < 0$. We define a vector point y^* in Y to be an <u>efficient</u> point in the set Y if there <u>does not</u> exist any other point y in Y such that $y \geq y^*$ i.e. $y_i \geq y_i^*$, i=1,2,...,n with at least one inequality strict.

Now let Y be a convex production set in R^n. If y^* is an efficient point in Y, then there exists a nonnegative vector p of prices $p \in R^n$ such that $p^T y^* \geq p^T y$ for all $y \in Y$. This result follows from the Minkowski Theorem. This price vector p may be interpreted as shadow prices and also related to the vector maximum problem of Kuhn-Tucker theory. Note two implications of the assumption of convexity of the production set Y above. One is that the firms have all all nonincreasing returns to scale and nonincreasing returns to each factor. Secondly, one could define a coefficient of efficiency, also called the coefficient of resource utilization by Debreu as ρ:

$$\rho = p^T y / p^T y^* \leq 1.0.$$

If for any feasible production process $y \in Y$, ρ is less than one, then it is inefficient at the prices p in the sense of vector efficiency. Note however that such prices may not be unique.

2.3 Nonlinear Maximization with Equality Constraints

One of the most frequently used maximization problems in applied economics is to maximize a function $y = f(x_1, x_2, \ldots, x_n)$ of n variables $x_1, x_2, \ldots, x_n$ subject to one constraint $g(x_1, x_2, \ldots, x_n) = 0$. For instance $f(x)$ may be the utility function of the consumer and the equality condition $g(x)$ may be the budget constraint. Alternatively, $y = f(x)$ may be the single output producible by the n inputs $x_1, x_2, \ldots, x_n$ and the constraint $g(x_1, \ldots, x_n) = 0$ may be the input cost constraint on the firm. We assume that $y = f(x)$ is at least twice differentiable.

First, consider the case when there is no constraint at all. Denote the partial derivatives of the function y by $f_i = \partial f / \partial x_i$, $i = 1, 2, \ldots, n$. The first order (necessary) condition for y to have extreme points, whether maxima or minima, yields

$$f_1 = f_1 = \ldots = f_n = 0 \qquad (8.1)$$

Any point satisfying this condition is called a critical point. For the second order (sufficient) condition for a maximum or minimum we need to define a Hessian determinant H of the function, which is based on the second partial derivatives

$$f_{ij} = \frac{\partial}{\partial x_i} \frac{\partial f}{\partial x_j} = \frac{\partial^2 f}{\partial x_i \partial x_j}, \qquad i, j = 1, 2, \ldots, n$$

where

$$H = \begin{vmatrix} f_{11} & f_{12} & \cdots & f_{1n} \\ f_{21} & f_{22} & \cdots & f_{2n} \\ \vdots & & & \\ f_{n1} & f_{n2} & \cdots & f_{nn} \end{vmatrix}$$

Once we have found the Hessian determinant one of the following three conditions must hold for any critical point satisfying the necessary condition (8.1):

(a) When H_1, H_2,..., H_n > 0 we have a minimum at the critical point, where

$$H_1 = |f_{11}| = f_{11}, \quad H_2 = \begin{vmatrix} f_{11} & f_{12} \\ f_{21} & f_{22} \end{vmatrix} \quad \text{and} \quad H_i \text{ is the i-th principal minor}$$

of the Hessian determinant.

(b) When H_1 < 0, H_2 > 0, H_3 < 0 and so on with alternating signs, we have a maximum at the critical point.

(c) When neither case (a) nor (b) holds, the test fails and we need to perturb the function in the neighborhood of the critical point in order to find out whether an extreme point exists.

2.3.1 Bordered Hessian

Now consider the case where we have one side condition $g(x_1, x_2, \ldots, x_n) = 0$ as an equality constraint subject to which $y = f(x)$ has to be maximized. Now we form the Lagrangean function $L = L(x_1, x_2, \ldots, x_n, \lambda) = f(x_1, x_2, \ldots, x_n) - \lambda g(x_1, x_2, \ldots, x_n)$ in $(n+1)$ variables $x_1, x_2, \ldots, x_n, \lambda$ and the necessary condition for an extremum is

$$\partial L / \partial x_i = 0, \quad i = 1, 2, \ldots, n$$
$$\partial L / \partial \lambda = 0. \tag{8.2}$$

This defines the critical points. To test whether any critical point satisfying (8.2) provides the maximum or the minimum, we must compute the bordered Hessian determinant $\bar{H}$ and its principal minors as follows:

$$\bar{H} = \begin{vmatrix} 0 & g_1 & g_2 & \cdots & g_n \\ g_1 & L_{11} & L_{12} & \cdots & L_{1n} \\ g_2 & L_{21} & L_{22} & \cdots & L_{2n} \\ \vdots & & & & \\ g_n & L_{n1} & L_{n2} & \cdots & L_{nn} \end{vmatrix}$$

where

$$\frac{\partial^2 L}{\partial x_i \partial x_j} = L_{ij} \text{ and } \frac{\partial g}{\partial x_i} = g_i.$$

The i-th bordered principal minor of a bordered Hessian determinant is denoted by $\bar{H}_i$ e.g.

$$\bar{H}_2 = \begin{vmatrix} 0 & g_1 & g_2 \\ g_1 & L_{11} & L_{12} \\ g_2 & L_{21} & L_{22} \end{vmatrix} \qquad \bar{H}_3 = \begin{vmatrix} 0 & g_1 & g_2 & g_3 \\ g_1 & L_{11} & L_{12} & L_{13} \\ g_2 & L_{21} & L_{22} & L_{23} \\ g_3 & L_{31} & L_{32} & L_{33} \end{vmatrix}$$

Note that the notation $\bar{H}_2$ means that we must take the Hessian determinant H_2 and place around it the appropriate border.

Now the second order (sufficient) condition says that if all the first order derivatives g_i of the constraint and all the second order derivatives L_{ij} of the Lagrangean function L exist at the critical point, then one of the following conditions must hold:

(a) When $\bar{H}_2$, $\bar{H}_3$,..., $\bar{H}_n$ < 0 we have a minimum at the critical point.
(b) When $\bar{H}_2$ > 0, $\bar{H}_3$ < 0, $\bar{H}_4$ > 0 and so on with alternating signs, we have a maximum at the critical point.
(c) When neither case (a) nor (b) holds, the test fails and we need to perturb the function in the neighborhood of the critical point to determine whether a constrained extremum exists.

As an example let $f = x_1^2 - 10x_2^2$ and $g = x_1 - x_2 - 18$. Then the first order condition applied to the Lagrangean $L = x_1^2 - 10x_2^2 - \lambda(x_1 - x_2 - 18)$ is

$$L_1 = 2x_1 - \lambda = 0, \quad L_2 = -20x_2 - \lambda = 0$$
$$L_\lambda = -(x_1 - x_2 - 18) = 0$$

from which we get the critical point $x_1 = 20$, $x_2 = 2$, $\lambda = 40$. At this point we evaluate the second order bordered principal minor for the second order sufficiency condition

$$\bar{H}_2 = \begin{vmatrix} 0 & g_1 & g_2 \\ g_1 & L_{11} & L_{12} \\ g_2 & L_{21} & L_{22} \end{vmatrix} = \begin{vmatrix} 0 & 1 & -1 \\ 1 & 2 & 0 \\ -1 & 0 & -20 \end{vmatrix} = 18.$$

Thus the point (20,2) defines a maximum point and not a minimum.

Two points may be noted about this derivation. One is that the bordered Hessian determinant is a symmetric determinant, since it is simply a Hessian determinant that is bordered by the first partial derivatives of the single constraint and zero. If there are more than one equality constraint, the above method may be readily extended. Secondly, if the objective function or the constraint does not have second partial derivatives, then the above sufficiency conditions would fail. We have then to think about using the Kuhn-Tucker theory of nonlinear programming.

2.4. A Critique of Optimization in Economic Theory

Three general types of criticisms have been made against the use of otpimizing (i.e. maximizing or minimizing) postulate in economic behavior. One is that this normative postulate fails to agree with actual observed behavior of consumers and producers. This is particularly true when the agents do not have all the information and there are definite costs of gathering and processing information. Herbert Simon has proposed a satisficing behavior as an alternative model of rational economic behavior. Thus instead of maximizing profit the entrepreneur may be satisfied if the profit level is at least some given level or higher. The second objection is that the optimizing postulate requires fine tuning or precise estimates of the parameters of the relevant functions, which are frequently unavailable in the real world. In situations of uncertainty, adaptive behavior is more realistic than the optimizing behavior. The former specifies a method of successive revision of decision rules and their updating as more information becomes available and some elements of caution, risk-aversion and learning are built into these rules. Thirdly, many economic situations present conflict of objectives between different agents as in noncooperative game theory and there are various possible solutions of such games depending on bargaining, negotiation and possible threats.

In spite of these objections the optimizing postulate used in normative models has found many valid applications in applied economics and econometrics. Recently in models known as rational expectations such normative behavior is frequently tested by econometric methods in macroeconomic models and the general equilibrium models of capital asset pricing.

PROBLEMS TWO

1. For a standard primal LP problem where x is the output-mix vector for a firm maximizing profits $z = \sum_{j=1}^{n} c_j x_j$ under the m resource constraints $\sum_{j=1}^{n} a_{ij} x_j \leq b_i$, $i=1,2,\ldots,m$ and $x_j \geq 0$, $j=1,2,\ldots,n$, derive the result that the dual of the dual is the primal problem.

2. (a) A stereo manufacturer makes three types of stereos: standard (y_1), quality (y_2) and deluxe (y_3). His profit margin from each is 15, 20 and 24 respectively. The standard model requires 3 hours for wiring and 1 hour for encasing. The quality model requires 1 hour for wiring and 5 hours for encasing. The deluxe model requires 3 hours for wiring and 2 hours for encasing. If 120 hours are avaialble for wiring and 60 hours for encasing, find the profit-maximizing output-mix.

   (b) Write down the dual of this problem and compute the optimal values of the dual variables. What is the interpretation of the dual in marginal cost terms?

3. Assume the quadratic programming problem

$$\max_{x_1, x_2} \quad z = (a_1 - b_1 x_1) x_1 + (a_2 - b_2 x_2) x_2$$

   subject to

$$\alpha_{11} x_1 + \alpha_{12} x_2 \leq \beta_1$$
$$x_1, x_2 \geq 0$$

   where $p_i = a_i - b_i x_i$ are the demand functions for the two products x_1, x_2.
   a) Write down the Kuhn-Tucker necessary and suficient conditions for an optimum, if

   $a_1 = 10$, $a_2 = 5$
   $b_1 = 1.0$, $b_2 = 0.5$
   $\alpha_{11} = 2$, $\alpha_{12} = 3$, $\beta_1 = 10$

   b) If b_1 and b_2 were zero, what would be the optimum solution?
   c) Write down the dual of the above quadratic program.

4. Solve the following LP problem
    $$\max z = 6x_1 + 2x_2$$
    $$\text{s.t. } 4x_1 + x_2 \leq 5, \quad 3x_1 + 2x_2 \leq 10$$
    $$x_1 \geq 0 \quad x_2 \geq 0$$

    and show its equivalent formulation in terms of two-person zero-sum game.
    a) How would the dual problem change if the nonnegativity constraints on x_1 and x_2 are dropped?
    b) Do you think that the optimal rules of peak load pricing apply in the LP model?

5. A scalar function $f(x) = f(x_1, x_2, \ldots, x_n)$ of the n-tuple vector x is called quasi-concave in a certain domain D, if for any two points $x(1)$, $x(2)$ in D the following holds for any scalar α, $0 \leq \alpha \leq 1$

    $$f(\alpha x(1) + (1-\alpha) x(2)) \geq \min [f(x(1)), f(x(2))]$$

    In other words if $x^o = \alpha x(1) + (1-\alpha) x(2)$ is the average of the two vector points $x(1)$, $x(2)$, then $f(x^o) \geq$ minimum of $(f(x(1)), f(x(2)))$. For strict quasi-concavity we replace the sign $\geq$ by $>$. Show that the function $f = x_1 + 2x_2 - 4x_1^2 - 2x_2^2$ in two nonnegative variables $x_1 \geq 0$, $x_2 \geq 0$ is quasi-concave.

6. A firm is seeking to determine the two input levels of labor (x_1) and capital (x_2) which minimize the total input cost $C = wx_1 + rx_2$ when the wage and rental rates are $w = 1$, $r = 2$ and the production constraint is given by $y = 120x_1^{0.5} x_2^{0.3}$ and the firm seeks to produce a level of output $\bar{y} = 1000$. Determine the optimal input levels x_1^o, x_2^o and perform the Hessian matrix test as a sufficiency condition.

7. For the consumer maximizing the utility function $u(x_1, x_2)$ in two goods subject to the budget constraint $p_1 x_1 + p_2 x_2 = m$ where p_i is the given price of good x_i, $i=1,2$ derive the optimal levels x_1^o, x_2^o of his purchase if the utility function is twice differentiable and concave. Then show that the bordered Hessian matrix is of the form

    $$\begin{bmatrix} 0 & p_1 & p_2 \\ p_1 & u_{11} & u_{12} \\ p_2 & u_{21} & u_{22} \end{bmatrix}$$

    where $u_{ij} = \partial^2 u(x_1, x_2)/\partial x_i \partial x_j$; $i, j = 1, 2$.

8. Let A be the Leontief input-output matrix

$$A = \begin{bmatrix} 0.2 & 0.7 \\ 0.4 & 0.1 \end{bmatrix}$$

with all positive elements. Then there exists a Frobenius eigenvalue λ^* solving the bilinear model

$$\max \lambda$$
$$\text{s.t.} \quad Ax \geq \lambda x \quad x \geq 0$$

Show that $\lambda^* = 0.55$. Derive the dual problem.

9. For the LP model in n-tuple vector x

$$\max z = c^T x$$
$$\text{s.t.} \quad Ax \leq b, \quad x \geq 0$$

show that the optimal pair (x^*, y^*) of primal dual variables satisfy the following conditions:

(a) $c^T x^* = b^T y^* = y^{*T} A x^*$

(b) $\dfrac{\partial z^*}{\partial b_i} = y_i^* \geq 0, \quad i=1,2,\ldots,m$

(c) $\partial z^*/\partial a_{ij} = -y_i^* x_j^*; \quad \begin{array}{l} i=1,2,\ldots,m \\ j=1,2,\ldots,n \end{array}$

where $z^* = c^T x^*$ is the maximand. These results are useful in testing the sensitivity of the maximand $z^* = c^T x^*$ with respect to variations in the parameters b_i, a_{ij}.

10. The payoff matrix of a two-person zero-sum game is given as

$$\begin{bmatrix} 4 & 3 \\ 1 & 7 \end{bmatrix}$$

Convert this game to an equivalent LP model and compute its optimal solution vector.

CHAPTER THREE
DYNAMIC SYSTEMS IN ECONOMICS

The Hicksian view of economic dynamics is that time must enter in a very essential way into the equations and functional relations describing economic behavior. Thus if y(t) is population in year t and it is growing at the rate of 2 percent per year, we can express the population rise by a <u>difference equation</u> in time as

$$y(t) = y(t-1) + 0.02\, y(t-1) = (1.02)\, y(t-1).$$

If the initial level of population, y(0) at t=0 is known, then one could predict the future level of population for any t > 0 by <u>successive iteration</u> as

t=1 $y(1) = a\, y(0),\ a=1.02$
t=2 $y(2) = a\, y(1) = a\, a y(0) = a^2 y(0)$
t=3 $y(3) = a\, y(2) = a\, a^2 y(0) = a^3 y(0)$
t=T $y(T) = a^T y(0).$

Hence for any t=s, $y(s) = a^s y(0)$, s=0,1,2,...,T. Instead of discrete time one may describe a growth variable in continuous time also. Thus let y_t be investment at time t, which grows at the constant instantaneous rate r = 0.05. Growth of y_t may be measured by the time derivative $\frac{dy}{dt}$ and one can specify the growth of investment by a <u>differential equation</u> in time as

$$\frac{dy}{dt} = ry = 0.05y$$

where the subscript t in y is omitted for convenience. Again if the initial value of y_t at t=0 is known as y_o, one could explicitly solve this differential equation for y as an explicit function of time t. The solution is

$$y_t = y_o\, e^{0.05t} = y_o\, e^{rt}$$

since by differentiating both sides we get back $dy/dt = ry_t = 0.05y$.

Thus difference and differential equations are examples of economic dynamics in the Hicksian sense, since we have the same variable y but with different time dimensions i.e. y(t), y(t-1) in difference equations and y_t, dy_t/dt in the differential case. In more complicated dynamic systems we may have mixed difference-differential equations or, partial differential equations which frequently arise in physics and engineering problems.

From an economic and social viewpoint one may mention four basic sources of dynamic phenomena in economics.

a) Lags in adjustment: many production situations involve time lags e.g. thus the supply of a crop cannot be increased before the next crop year even though demand suddenly rises due to export. Other important lags in economics are the gestation lags when plant capacity is increased over a period of time, lags between income and expenditure and lags between expected earnings and realized earnings. One way to model such lags is through the difference equations.

b) Incidence of time: the decisions of economic agents i.e. consumers and producers are significantly influenced by their planning horizon and life expectancy. Decisions which are appropriate in the short run may not be so in the long run. In econometric analysis of time series, which deals with economic variables changing over time a distinction is made between the cyclical analysis (short run) and the trend analysis (long run). Thus monetary variables like money supply or interest rates are important subjects for short run economic modeling, whereas real factors like population growth, real capital formation and technology growth are matters for secular analysis or long run growth of an economy. Thus the sources of short run fluctuations in economic activity may be different from the long run instability.

c) Disequilibrium analysis: the demand-supply equilibrium of a competitive market can be understood better by dynamic analysis, when due to some exogenous shocks the equilibrium is disturbed. The capacity of a market to adjust to a new equilibrium is a stabilizing property, which may be useful to policy makers who intend to stabilize such markets e.g. international commodity markets. Other equilibrium concepts in economics in general equilibrium models such as Leontief's input-output model may be similarly analyzed for their behavior under disequilibrium. Recently, the equilibrium concepts in economics have come under attack by those who propose ante-equilibrium or, some form of disequilibrium as a probable model of economic behavior. Many recent econometric models incorporate some form of disequilibrium behavior to allow for the fact that the economic agents do not have most of the time all the information to adopt the equilibrium behavior which is closely related to normative and optimizing decision rules.

d) Expectational variables: the rational expectations model of Muth, extensively utilized in recent monetary policy debates puts strong emphasis on agents' anticipations profoundly influencing the dynamic processes between equilibrium points. There is not much disagreement among modern economists that in the real world, equilibrium in various markets (e.g. labor, goods and financial markets) is a stochastic equilibrium i.e an equilibrium affected by stochastic factors like anticipations or expectations and that in such an equilibrium the average actual and anticipated prices should be equal. However there is considerable disagreement among economists today concerning the ability of the competitive market to effectively anticipate changes in prices in an unbiased manner, when the equilibrium price is changing. The Muth rational expectation (RE) hypothesis states that in the aggregate, the anticipated or expectational price is an unbiased predictor of the actual price i.e. on the average the two are equal. For instance let $p_s^*(t)$ be the price at time t that was anticipated at an earlier date s, s < t when spending and supply decisions were made by the economic agents; let E denote the arithmetic average, p(t) be the realized (or actual) price at time t and e(t) the forecast error and I(s) be the information set containing all relevant information upto time s. Then the RE hypothesis is

$$p_s^*(t) = E\{p(t)|I(s)\}$$
$$e(t) = p(t) - p_s^*(t); \qquad E\{e(t)\} = 0, \text{ all } t.$$

Since the expectational (or anticipated) variable $p_s^*(t)$ is not directly observable from actual price data, various dynamic models are formulated through suitable proxy variables or alternative assumptions.

3.1. Linear Difference Equation Systems

Difference equations are more realistic for most economic systems because the economic time series data are usually in discrete units of time. Hence this type of dynamic system is easier to estimate by econometric methods against empirical data.

Since the difference equation $y(t+1) = ay(t)$ can also be written in terms of the finite increment $\Delta y(t) = y(t+1) - y(t)$ as $\Delta y(t) = (a-1)y(t)$, one could write for the ratio $\Delta y(t)/\Delta t$ where $\Delta t = (t+1) - t = 1$ the continuous time approximation dy_t/dt and obtain a differential equation approximation $dy/dt = (a-1)y$. Both difference and differential equations in linear form can be described in a common notation by using the operator notation. Let D be the operator such that $Dy(t) = y(t+1)$, it is then called a forward difference operator. But if D is a differential operator then $Dy_t = dy/dt$. The degree of the operator D denotes the <u>order</u> of the linear dynamic equation. Thus the difference equation

$$y(t+2) + 2y(t+1) - 4y(t) + 3 = 0$$

can be written in terms of the forward difference operator D

$$D^2 y(t) + 2Dy(t) - 4y(t) + 3 = 0.$$

Note that $D^2 y(t) = D(Dy(t)) = D(y(t+1)) = y(t+2)$, since the highest degree of D is two, this is a second order difference equation. This equation is also nonhomogeneous, since it has a term (+3) independent of $y(t)$. On using the coefficients a_i, an n-th order linear differential or difference system can similarly written as

$$(a_n D^n + a_{n-1} D^{n-1} + \ldots + a_2 D^2 + a_1 D + a_0) y = b$$

where a_n is not zero, D may be the difference or differential operator and b is the nonhomogeneous term. If b is zero, we have an n-th order linear difference or differential equation in the dated variable y.

3.1.1 First Order Equation

The linear first order difference equation with constant coefficients and a constant nonhomogeneous term is perhaps the easiest to solve by successive iteration. However since we want to extend the solution procedure to the n-th order case for $n \geq 2$, we use a more systematic method, called the method of superposition. By this method we solve the linear nonhomogeneous difference equation

$$y(t+1) - ay(t) = b \qquad (1.1)$$

with constant nonzero coefficients a,b in three steps. First, we solve the homogeneous part of the equation i.e. $y(t+1) - ay(t) = 0$ by setting b to zero. The solution is $y(t) = Aa^t$ where A is a constant to be determined (we cannot set A equal to the initial value $y(0)$, since we have not yet incorporated the nonhomogeneous part). The second step is to obtain a _particular_ solution of the complete nonhomogeneous equation. If the nonhomogeneous term is constant as in this case, then the particular solution is given by an equilibrium or steady-state value y^* say i.e. a value such that if $y(t)$ were fixed at y^* then it would stay there forever. In other words, $y(t+1) = y^*$ and $y(t) = y^*$ for all t. Hence we obtain from the complete equation

$$y^* - ay^* = b, \text{ or } y^* = b/(1-a), a \neq 1.$$

Note that we need $a \neq 1$ for y^* to exist. The third step is to add the two solutions to obtain the complete general solution for any t as

$$y(t) = A a^t + y^* = A a^t + \frac{b}{1-a} \qquad (1.2)$$

where we determine now the constant A by the initial condition $y(0)$ of $y(t)$ at $t=0$ from (1.2) by substitution

$$y(0) = A\,a^0 + \frac{b}{1-a} = A + \frac{b}{1-a} \quad \text{since } a^0 = 1$$

or

$$A = y(0) - \frac{b}{1-a}.$$

Thus (1.2) becomes

$$y(t) = [y(0) - \frac{b}{1-a}]\,a^t + \frac{b}{1-a}. \tag{1.3}$$

In case $a=1$, the particular solution has to be tried in a time-dependent form $y^*(t) = kt$ which implies $y^*(t+1) = k(t+1)$ and therefore we get from (1.1)

$$y^*(t+1) - a y^*(t) = k(t+1) - akt = b$$

or

$$k = \frac{b}{t+1-at} = b, \text{ since } a=1 \qquad y^*(t) = kt = bt.$$

The complete general solution is therefore

$$\begin{aligned} y(t) &= A\,a^t + bt \\ &= y(0) + bt \qquad \text{for } a=1. \end{aligned} \tag{1.4}$$

<u>Ex 1</u> (Cobweb model of price fluctuations)

As a simple example we consider the market model of agricultural prices of agricultural crops like corn, where demand $d(t)$ depends on current price $p(t)$ but supply $s(t)$ on lagged price $p(t-1)$

(demand) $\quad d(t) = a_1 - b_1 p(t) \qquad a_1, b_1 > 0$
(supply) $\quad s(t) = -a_2 + b_2 p(t-1) \qquad a_2, b_2 > 0$

market clearing condition $d(t) = s(t)$, all t.

By the market clearing condition, the prices $p(t)$ must satisfy the first-order linear difference equation

$$p(t) = c_1 + c_2\,p(t-1), \quad c_1 = \frac{a_1 + a_2}{b_1}, \quad c_2 = -b_2/b_1$$

which can also be written as

$$p(t+1) = c_1 + c_2\,p(t), \text{ since } t \text{ is arbitrary.}$$

By analogy with (1.3), the complete general solution of the equilibrium price path is

$$p(t) = [p(0) - \frac{c_1}{1 - c_2}](c_2)^t + \frac{c_1}{1 - c_2}$$

$$= [p_o - \frac{a_1 + a_2}{b_1 + b_2}](-\frac{b_2}{b_1})^t + \frac{a_1 + a_2}{b_1 + b_2} \qquad (1.5)$$

where p_o is used for the initial price $p(0)$. Several features of this price path in (1.5) may be noted. First, the price path solution may be written here in terms of the equilibrium or steady-state value p^* as

$$p(t) = (p_o - p^*)(-\frac{b_2}{b_1})^t + p^*.$$

Hence if the ratio b_2/b_1 is less than one, then the limit of $p(t)$ converges to p^* as $t \to \infty$, since $(-b_2/b_1)^t$ becomes zero for large t. Otherwise it does not converge. Clearly, in the convergent case p^* defines the static equilibrium where the demand and supply curves intersect. Secondly, the transient part of the solution (i.e. the solution of the homogeneous part) involving the term $c_2^t = (-b_2/b_1)^t$ shows oscillatory behavior i.e. c_2^t is positive for t even but negative for t odd. There can be three types of oscillations depending on the value of the slope coefficients i.e. the oscillation will be <u>explosive</u> for $b_2 > b_1$, <u>damped</u> if $b_1 > b_2$ and uniform if $b_2 = b_1$. Clearly if a stabilizing tendency of $p(t)$ converging to the equilibrium price p^* is considered a desirable objective, then the demand slope (b_1) must exceed the supply slope (b_2). Therefore any policy measure like the buffer fund policy which tends to increase the demand slope and/or decrease the supply slope thereby reducing the ratio b_2/b_1 below one would tend to contribute to the stabilizing trend of price movement. Also the oscillation in price movement could also be reduced. Thus let $g(t)$ be the buffer fund policy where

where $\qquad d(t) = s(t) + g(t)$
$\qquad\qquad g(t) = k_1 p(t) - k_2 p(t-1) \qquad k_1, k_2 > 0.$

In other words the buffer fund policy increases the slope of demand and reduces the slope of supply. The reduced form equation now becomes

$$p(t) = \frac{k_2 - b_2}{k_1 + b_1} p(t-1) + \frac{a_1 + a_2}{k_1 + b_1}.$$

The complete general solution is

$$p(t) = (p_o - p^*)(\frac{k_2 - b_2}{k_1 + b_1})^t + p^*$$

with $\quad p^* = (a_1 + a_2)/(k_1 + b_1)$.

Clearly if k_1, k_2 are such that $k_2 > b_2$ and $(k_1 + b_1) > (k_2 - b_2)$, then the path of convergence of $p(t)$ to the steady-state equilibrium value p^* is assured and the path is nonoscillatory.

<u>Ex 2</u> (Walrasian model of price adjustment)

The Walrasian model of price adjustment for a competitive market model with no time-lags in supply or demand assumes that the presence of excess demand (excess supply) would tend to increase (decrease) the market price i.e.

$$\Delta p(t) = p(t+1) - p(t) = k (d(t) - s(t)) \tag{1.6}$$

where
$\quad k =$ positive scalar = speed of adjustment,
$\quad e(t) = d(t) - s(t) =$ excess demand if $e(t) > 0$
$\quad$ and excess supply if $e(t) < 0$
$\quad d(t) = a_1 - b_1 p(t) \qquad a_1, b_1 > 0$
$\quad s(t) = -a_2 + b_2 p(t) \qquad a_2, b_2 > 0.$

The equilibrium or market clearing price is

$$p^* = \frac{a_1 + a_2}{b_1 + b_2} \text{ where } d(t) = s(t)$$

and the Walrasian adjustment equation (1.6) has the general solution

$$p(t) = (p_0 - p^*)[1 - k(b_1 + b_2)]^t + p^*.$$

The adjustment path $p(t)$ converges to p^* if $|c| < 1$, where $c = 1 - k(b_1 + b_2)$, otherwise it diverges. If $k(b_1 + b_2) > 1$ but $|c| < 1$ then the path is convergent but oscillatory. Again by a properly designed buffer fund policy the stabilizing tendency of the price adjustment path could be improved.

3.1.2 Second Order Equation

On using the forward difference operator D as $Dy(t) = y(t+1)$, a second order linear difference equation with constant coefficients a_1, a_2 and a constant non-homogeneous term b can be written as

$$(D^2 + a_1 D + a_2) y(t) = b. \tag{2.1}$$

As a particular or steady-state solution we may try the constant level $y^* = y(t-1) = y(t-2) = y(t)$ for all t

$$y^* = b/(1 + a_1 + a_2) \tag{2.2}$$

provided $a_1 + a_2 \neq -1$. In case $a_1 + a_2 = -1$, the trial solution $y^* =$ constant fails and as before we try $y^*(t) = kt$ and obtain the particular solution as

$$y^*(t+2) + a_1 y^*(t+1) + a_2 y^*(t) = b$$
or
$$k(t+2) + a_1 k(t+1) + a_2 kt = b \tag{2.3}$$

or, $k = b/[(1 + a_1 + a_2) t + a_1 + 2] = \dfrac{b}{a_1 + 2}$

since $a_1 + a_2 = -1$.

The homogeneous solution is to be found by trying the solution as $y(t) = A\lambda^t$ from which one gets

$y(t+1) = A\lambda^{t+1}$, $y(t+2) = A\lambda^{t+2}$ and hence

$$(D^2 + a_1 D + a_2) y(t) = 0$$

becomes

$$A\lambda^{t+2} + a_1 A\lambda^{t+1} + a_2 A\lambda^t = 0$$
or,
$$A\lambda^t (\lambda^2 + a_1 \lambda + a_2) = 0 \tag{2.4}$$

where A which is to be determined by the initial conditions is still arbitrary. It is clear that if $A\lambda^t$ is a solution of the second order homogeneous equation, then we must have from (2.4)

$$\lambda^2 + a_1 \lambda + a_2 = 0 \tag{2.5}$$

this is called the <u>characteristic equation</u> or the eigenvalue equation in eigenvalue λ. (Note that in the first order case (1.1) the characteristic equation was $\lambda - a = 0$ i.e. $\lambda = a$). Note that the characteristic equation (2.5) denoted as $g(\lambda_1, \lambda_2) = \lambda^2 + a_1 \lambda + a_2 = 0$ is quadratic i.e. of order 2. In general, an n-th order linear homogeneous difference equation with constant coefficients has a characteristic equation of order n i.e. $g(\lambda) = g(\lambda_1, \lambda_2, \ldots, \lambda_n) = 0$ and the n values of $\lambda_1, \lambda_2, \ldots, \lambda_n$ are also called n characteristic roots or eigenvalues. The two characteristic roots of (2.5) associated with the second order system (2.1) in its homogeneous form are

$$\lambda_1, \lambda_2 = (-a_1 \pm \sqrt{a_1^2 - 4a_2})/2. \tag{2.6}$$

Again by the superposition theorem the parts λ_1^t and λ_2^t are to be linearly combined each with its arbitrary constant A_1 and A_2, so that the homogeneous solution is

$$y(t) = A_1 \lambda_1^t + A_2 \lambda_2^t. \tag{2.7}$$

Here the two arbitrary constants A_1, A_2 must be independent, otherwise the linear combination of the two terms λ_1^t, λ_2^t would not define a valid combination e.g. if A_1 is set to zero, λ_1^t does not have any role in the solution (2.7). Four important cases of the two roots in (2.6) must be clearly distinguished.

<u>Case 1</u> (distinct real roots): when $a_1^2 - 4a_2 > 0$, the two roots are real and distinct. The complete solution in this case is

$$y(t) = A_1 \lambda_1^t + A_2 \lambda_2^t + y^*. \tag{2.7}$$

where A_1 and A_2 must be determined from the two initial conditions e.g. $y(0) = y_o$ and $y(1) = y_1$

$$y(0) = y_o = A_1 \lambda_1^0 + A_2 \lambda_2^0 + y^* = A_1 + A_2 + y^*$$
$$y(1) = y_1 = A_1 \lambda_1 + A_2 \lambda_2 + y^*$$

or,

$$A_1 = \frac{(y_o - y^*) \lambda_2 - (y_1 - y^*)}{\lambda_2 - \lambda_1}, \quad A_2 = \frac{y_1 - y^* - \lambda_1 (y_o - y^*)}{\lambda_2 - \lambda_1}$$

$$y^* = (1 + a_1 + a_2)^{-1} b, \text{ assuming } a_1 + a_2 \neq 1.$$

For example the equation $2y(t+2) - 4y(t+1) + 3y(t) = 5$ with the two initial conditions $y(0) = y_o = 1$ and $y(1) = y_1 = 2$ has the general solution

$$y(t) = -(3/2)^t - 3(1/2)^t + 5.$$

<u>Case 2</u> (equal real roots): $\lambda_1 = \lambda_2$ when $a_1^2 = 4a_2$ and in this case the linear combination of the two solution components λ_1^t, λ_2^t collapse into a single terms

$$A_1 \lambda_1^t + A_2 \lambda_2^t = (A_1 + A_2) \lambda_1^t = A_3 \lambda_1^t.$$

But by superposition theorem we need two constants and not one. To supply the missing component A_4 say we must note however that it must be independent of A_3. By applying the same old trick as in the particular solution we use two independent components as $A_3 \lambda_1^t$ and $A_4 t \lambda_1^t$ that is, $y(t) = A_3 \lambda_1^t + A_4 t \lambda_1^t$ as the solution of the homogeneous equation. The second component $(A_4 t \lambda_1^t)$ is independent of the first $(A_3 \lambda_1^t)$, since we cannot obtain the first by multiplying the second by any constant.

Hence in this case the complete solution is

$$y(t) = A_3 \lambda_1^t + A_4 t \lambda_1^t + y^* \tag{2.8}$$

where A_3 and A_4 are determined by the two initial conditions e.g. $A_3 = y_0 - y^*$ and $A_4 = [y_1 - y^* - \lambda_1(y_0 - y^*)]/\lambda_1$.

<u>Case 3</u> (real and reciprocal roots): when $\lambda_2 = 1/\lambda_1$ and both roots are real, the solution of the homogeneous equation is of the form $y(t) = A_1 \lambda_1^t + A_2 (1/\lambda_1)^t$ where the two independent constants A_1, A_2 are determined by initial conditions. In this case if λ_1 is a positive fraction, one component $A_1 \lambda_1^t$ decreases to zero as t goes to infinity but the other component $A_2(1/\lambda_1)^t$ explodes to infinity since $1/\lambda_1$ exceeds one. This behavior is called saddle point, which occurs in dynamic optimization with linear quadratic models to be discussed in the next chapter. In this case for $y(t)$ to converge to the steady-state constant level y^*, we need to have such initial or terminal conditions that A_2 can be set to zero and this is precisely done by the so-called transversality condition in optimal control theory.

<u>Case 4</u> (complex or imaginary roots): when $a_1^2 < 4a_2$, we have the two roots λ_1, λ_2 conjugate complex. They are of the form

$$\lambda_1 = \alpha + \beta i \text{ and } \lambda_2 = \alpha - \beta i, \quad i = +\sqrt{-1}$$

where α, β are real numbers and i is an imaginary number. Here $\alpha = -a_1/2$, $\beta = \sqrt{(4a_2 - a_1^2)}/2$. By De Moivre's theorem the complex number $(\alpha \pm \beta i)^t$ can be expressed in trigonometric terms as

$$(\alpha + \beta i)^t = R^t (\cos\theta t \pm i \sin\theta t)$$

where

$$R = \sqrt{\alpha^2 + \beta^2} = \left(\frac{a_1^2 + 4a_2 - a_1^2}{4}\right)^{\frac{1}{2}} = \sqrt{a_2}$$

and θ is the radian measure of the angle in the interval $[0, 2\pi]$ which satisfies the conditions

$$\cos\theta = \alpha/R = \frac{-a_1}{2\sqrt{a_2}}, \quad \sin\theta = \frac{\beta}{R} = \left(\frac{4a_2 - a_1^2}{4a_2}\right)^{1/2}.$$

Hence the general solution

$$y(t) = A_1(\alpha + \beta i)^t + A_2(\alpha - \beta i)^t + y^*$$

can be conveniently written as

$$y(t) = y^* + A_1 R^t(\cos\theta t + i\sin\theta t) + A_2 R^t(\cos\theta t - i\sin\theta t)$$
$$= y^* + R^t(A_3\cos\theta t + A_4\sin\theta t) \qquad (2.9)$$

where $A_3 = A_1 + A_2$, $A_4 = (A_1 - A_2)i$.

As a simple example, take the equation $y(t+2) + 0.25\,y(t) = 6$. We have $a_2 = 1/4$, $a_1 = 0$ and therefore $\alpha = 0$, $\beta = 1/2$, $R = 1/2$, $\cos\theta = 0$, $\sin\theta = 1$. Hence $\theta = \pi/2$. The general solution is

$$y(t) = (1/2)^t \left(A_3 \cos\frac{\pi}{2}t + A_4 \sin\frac{\pi}{2}t\right) + 6$$

where A_3, A_4 are determined by the two initial conditions y_0 and y_1.

The complex or imaginary roots contribute fluctuations of a periodic nature i.e. the cycles repeat after certain periods - just like ocean or sound waves.

The convergence of the time-path of solution $y(t)$ to y^*, the steady-state constant equilibrium level depends in the second order case on the nature of the two characteristic roots λ_1, λ_2. The general rule is that each characteristic root must be less than unity in absolute value for the path to converge. This means that for cases 1-3 of real roots, the absolute value of each root $|\lambda_i|$ must be less than one. In case 2 of equal real roots, it may appear from (2.8), that the second term on the right-hand side i.e. $A_4 t \lambda_1^t$ might explode as $t \to \infty$, even if $|\lambda_1| < 1$, but the nonlinear part λ_1^t goes to zero faster, hence $(t\lambda_1^t)$ goes to zero as $t \to \infty$ provided $|\lambda_1| < 1$. For the complex root (case 4) case, the absolute value of the complex root λ_1, $\lambda_2 = \alpha \pm \beta i$ is given by $R = +\sqrt{(\alpha^2 + \beta^2)}$ and it is apparent from the form of the general solution (2.9), that the term

$$R^t(A_3\cos\theta t + A_4\sin\theta t) \to \text{zero as } t \to \infty, \text{ if } R < 1$$

hence $y(t) \to y^*$ as $t \to \infty$.

3.1.3 n-th Order Equation

For the n-th order linear difference equation $(a_0 D^n + a_1 D^{n-1} + a_2 D^{n-2} + \ldots + a_{n-1} D + a_n) y(t) = b$ with constant coefficients $a_0, a_1, \ldots, a_n$ and a constant non-homogeneous term b, the characteristic equation for the homogeneous part

$$g(\lambda) = a_0 \lambda^n + a_1 \lambda^{n-1} + \ldots + a_{n-1} \lambda + a_n = 0 \tag{3.1}$$

is an n-th degree polynomial equation and hence has n roots or solutions denoted by λ_i say with $i=1,2,\ldots,n$. If the n roots are all real and distinct, then the homogeneous solution can be written as

$$y(t) = \sum_{i=1}^{n} A_i \lambda_i^t \tag{3.2}$$

where the n independent constants are determined by n given initial conditions. The complete solution is $y(t) = \sum_{i=1}^{n} A_i \lambda_i^t + y^*$. If each root is less than one in absolute value ($|\lambda_i| < 1$, $i=1,2,\ldots,n$) then the $y(t)$ time-path converges to its constant steady-state equilibrium y^* i.e. the path is said to be stable. If any one of λ_i is greater than one in absolute value with its corresponding A_i nonzero, the path is not stable i.e. it diverges from y^* once it is disturbed.

Other cases we have discussed before in case (b) can be easily analyzed now as special versions. For example, if there are three repeated real roots say $\lambda_1 = \lambda_2 = \lambda_3$ then the first three terms on the right-hand side of (3.2) would be transformed as

$$A_1 \lambda_1^t + A_2 t \lambda_1^t + A_3 t^2 \lambda_1^t$$

Again if there is a pair of complex conjugate roots say λ_1 and λ_2 then the first two terms on the right-hand side of (3.2) would be

$$R^t (A_1 \cos\theta t + A_2 \sin\theta t)$$

and for stability or convergence we need $R < 1$.

In general for the n-th order system (3.1), the stability can be tested by the <u>Schur Theorem</u>, which states that the following determinants must be all positive for a stable (or convergent) system

$$\Delta_1 = \begin{vmatrix} a_0 & a_n \\ \hline a_n & a_0 \end{vmatrix} \qquad \Delta_2 = \begin{vmatrix} a_0 & 0 & a_n & a_{n-1} \\ a_1 & a_0 & 0 & a_n \\ \hline a_n & 0 & a_0 & a_1 \\ a_{n-1} & a_n & 0 & a_0 \end{vmatrix}$$

and so on upto Δ_n. In other words $\Delta_i > 0$, for i=1,2,...,n is a necessary and sufficient condition for stability or convergence. The construction of these determinants $\Delta_1, \Delta_2, \Delta_3, \ldots, \Delta_4$ follows a simple procedure. This is explained by the partition lines which divide the determinant into four areas. Each area of the j-th determinant Δ_j always consists of a j by j subdeterminant. The upper left area has a_0 alone in the diagonal, zeros above the diagonal and progressively higher subscripts for the successive coefficients in each column below the diagonal elements. The lower-right subdeterminants is just the transpose of the subdeterminant in the upper-left area. For example with n=3

$$\Delta_3 = \left| \begin{array}{ccc|ccc} a_0 & 0 & 0 & a_3 & a_2 & a_1 \\ a_1 & a_0 & 0 & 0 & a_3 & a_2 \\ a_2 & a_1 & a_0 & 0 & 0 & a_3 \\ \hline a_3 & 0 & 0 & a_0 & a_1 & a_2 \\ a_2 & a_3 & 0 & 0 & a_0 & a_1 \\ a_1 & a_2 & a_3 & 0 & 0 & a_0 \end{array} \right|$$

For the upper-right area, we place the a_n (n=3) coefficients above in the diagonal, with zeros below the diagonal and progressively lower subscripts for the successive coefficients as we go up each column from the diagonal. The lower-left subdeterminant is just the transpose of the one in the upper-right area. For example the equation $y(t+2) - 3y(t+1) + 2y(t) = 0$ with n=2 has the two Schur determinants ($a_0 = 1$, $a_1 = -3$, $a_2 = 2$):

$$\Delta_1 = \left| \begin{array}{cc} a_0 & a_2 \\ a_2 & a_0 \end{array} \right| = \left| \begin{array}{cc} 1 & 2 \\ 2 & 1 \end{array} \right| = -3, \quad \Delta_2 = -10$$

Clearly it violates the Schur Theorem on stability. Hence the solution is not stable.

3.1.4 Time-Dependent Nonhomogeneous Term

When the nonhomogeneous term $b(t)$ is dependent on time t we need some care in specifying the particular solution now denoted as $y_p(t)$ and also the solution of the homogeneous part now denoted by $y_H(t)$. We follow the method of undetermined coefficients in four steps as follows.

Step 1. Find the homogeneous solution $y_H(t)$.

Step 2. Construct the entire family for every term in the nonhomogeneous function $b(t)$. For example, if $b(t) = c_1 a^t + c_2 t^h$ where a, h, c_1, c_2 are constants and h is integer, we list the family members for a^t, which is of course a^t only and the members for t^h, which is $t^h, t^{h-1}, \ldots, t$ and 1.

Step 3. Check the homogeneous solution to see if any member of any finite family has the same functional form as any term in y(t). If so, every member of the family whose member corresponds to a term in $y_H(t)$ is multiplied by the <u>smallest</u> <u>integer</u> <u>power</u> of t such that no member of the family will then have the same functional form as any term in $y_H(t)$. To illustrate let us assume $b(t) = c_1 a^t + c_2 t^h$ as in Step 2 and let $y_H(t) = c_3 t + c_4$ be the homogeneous solution, which it could if its characteristic equation had a double root of one. Two members of the family of t^h, namely, t and the constant one, have the same functional form as a term in $y_H(t)$. Thus every member of the family of t^h must be multiplied by some power of t. Simple multiplication by t would not be sufficient since the constant member of the family would then become t which still corresponds to the t term in $y_H(t)$. Hence multiplication by t^2 is required to fully remove the correspondence between the family and all terms in $y_H(t)$. The resulting family terms therefore become a^t, which remains unchanged since it is in a different family, and $t^{h+2}, t^{h+1}, \ldots, t^3$ and t^2. No t or constant term remains.

Step 4. Construct $y_p(t)$ as a linear combination of all family members, modified where appropriate by multiplication by powers of t in step 3 for the families corresponding to all terms in the nonhomogeneous function b(t). For our example above in steps 2 and 3, the assumed form of $y_p(t)$ would be

$$y_p(t) = Aa^t + Bt^{h+2} + Ct^{h+1} + \ldots + Mt^3 + Nt^2$$

where A through N are constants to be determined.

Step 5. Determine the combinatorial coefficients or constants A through N in step 4 above in $y_p(t)$ such that the original difference equation is identically satisfied when the assumed $y_p(t)$ is substituted for y(t) in the complete equation.

Consider a second order example

$$y(t+2) - 3y(t+1) + 2y(t) = 4^t + 3t^2. \tag{3.3}$$

The homogeneous equation $(D^2 - 3D + 2)y(t) = 0$ yields the characteristic equation $\lambda^2 - 3\lambda + 2 = 0$ which has two roots $\lambda_1 = 2$, $\lambda_2 = 1$. Thus the homogeneous solution is

$$y_H(t) = c_1 2^t + c_2 \qquad c_1, c_2 \text{ are arbitrary constants to be determined by initial conditions.}$$

Here the nonhomogeneous function is the sum of two different functions of t, namely 4^t and t^2. Clearly 4^t is a special case of a^t which is a one-member family. Thus 4^t is the entire family. The family associated with t^2 contains t^2, t and 1. We have to check for corresponding functions of $y_H(t)$. Although 2^t appears in $y_H(t)$, it is not the same function as 4^t, so no correspondence exists for 4^t. The constant term 1, however corresponds to the constant term in $y_H(t)$. Therefore the entire family of t^2, t and 1 must be multiplied by t to obtain t^3, t and t respectively. This clears the correspondence and yields the assumed form of the particular solution

$$y_p(t) = A4^t + Bt^3 + Ct^2 + Dt$$

where we have to determine the constants A, B, C and D. By substitution, we get

$$\begin{aligned}y(t+1) &= A4^{t+1} + B(t+1)^3 + C(t+1)^2 + D(t+1) \\ &= 4A \cdot 4^t + Bt^3 + (3B + C)t^2 \\ &\quad + (3B + 2C + D)t + (B + C + D)\end{aligned}$$

and also,

$$\begin{aligned}y(t+2) &= 16A \cdot 4^t + Bt^3 + (6B + C)t^2 \\ &\quad + (12B + 4C + D)t + (8B + 4C + 2D).\end{aligned}$$

On substituting these expressions in the original difference equation (3.3) and regrouping terms in like functions of t we obtain

$$\begin{aligned}&6A \cdot 4^t + 0 \cdot t^3 - 3B\, t^2 + (3B - 2C)t + (5B + C - D) \\ &= 4^t + 3t^2.\end{aligned}$$

For this equation to be satisfied for all values of t, the coefficients of each function of t on each side of the equation must be equal. Hence

$6A = 1$, (equating coefficients of 4^t), or $A = 1/6$
$-3B = 3$ (equating coefficients of t^2), or $B = -1$
$3B - 2C = 0$ (equating for t), or $C = 3/2$
and $(5B + C - D) = -5 - 3/2 - D = 0$, or $D = -7/2$.

Hence the particular solution is

$$y_p(t) = (1/6)4^t - t^3 + (3/2)t^2 - (7/2)t.$$

The complete solution is

$$y(t) = y_H(t) + y_p(t) = (1/6)4^t - t^3 + (3/2)t^2 - (7/2)t + c_1 2^t + c_2$$

where the independent constants c_1, c_2 are to be determined by two initial conditions say y_0 and y_1 i.e.

$$y(0) = y_0 = 1/6 + c_1 + c_2$$
$$y(1) = y_1 = 2/3 - 1 + 3/2 - 7/2 + 2c_1 + c_2$$

or, $c_1 = 21/6$ and $c_2 = -8/3$, when $y_0 = 1$, $y_1 = 2$.

3.1.5 Simultaneous Equations

By the superposition theorem, it is always possible to transform an n-th order linear difference (or differential) equation in <u>one</u> variable into n simultaneous first order linear difference (or differential) equations. For example consider the difference equation $y(t+2) + a_1 y(t+1) + a_2 y(t) = b$ with constant coefficients a_1, a_2, b. Introduce a new variable $z(t) = y(t+1)$ and express the single second order equation as two first order equations as follows

$$z(t+1) + a_1 z(t) + a_2 y(t) = b$$
$$y(t+1) - z(t) = 0 \qquad (3.4)$$

Conversely, given two linear first order equations in (3.4) one can get back one second order linear difference equation

$$y(t+2) + a_1 y(t+1) + a_2 y(t) = b$$

by eliminating $z(t)$ from the two equations of (3.4), which define a simultaneous or jointly determined dynamic system.

A second way of solving a set of linear difference equations is by using a matrix method. As an example we take a simple cobweb model in two crops (corn and wheat say) for which the demand, supply and prices are $d_i(t)$, $s_i(t)$, $p_i(t)$ with i=1,2, and the supply is lagged by one time unit

$$d_1(t) = a_{11} p_1(t) + b_1$$
$$s_1(t) = a_{21} p_2(t-1) + b_2$$
$$d_2(t) = a_{31} p_2(t) + b_3$$

$$s_2(t) = a_{41}p_2(t-1) + a_{42}p_1(t-1) + b_4$$
$$d_i(t) = s_i(t), \text{ at all } t, i=1,2 \text{ (market clearing)}.$$

On using the market-clearing condition we can write the simultaneous linear system in terms of vectors p(t) as

$$\begin{pmatrix} p_1(t) \\ p_2(t) \end{pmatrix} = \begin{pmatrix} \alpha_{11} & \alpha_{12} \\ \alpha_{21} & \alpha_{22} \end{pmatrix} \begin{pmatrix} p_1(t-1) \\ p_2(t-1) \end{pmatrix} + \begin{pmatrix} \beta_1 \\ \beta_2 \end{pmatrix}$$

where $\alpha_{11} = a_{21}/a_{11}$, $\alpha_{12} = 0$, $\alpha_{21} = a_{42}/a_{31}$, $\alpha_{22} = a_{41}/a_{31}$, $\beta_1 = (b_2-b_1)/a_{11}$ and $\beta_2 = (b_4-b_3)/a_{31}$ or,

$$p(t) = A\, p(t-1) + b \qquad (3.5)$$

where p(t), b are 2-tuple column vectors with typical elements $p_i(t)$ and β_i respectively and A is a 2 by 2 coefficient matrix associated with the lagged price vector p(t-1) with a typical element α_{ij} (i,j=1,2). Assuming b to be a constant vector we determine the particular solution as a vector p* of steady-state or equilibrium values by setting p(t) = p* = p(t-1) for all t

$$p^* = A\, p^* + b$$

or,

$$p^* = (I-A)^{-1} b, \qquad I = \begin{bmatrix} 1 & 0 \\ 0 & 1 \end{bmatrix} \qquad (3.6)$$

assuming that the matrix (I-A) is nonsingular so that its inverse $(I-A)^{-1}$ exists. The homogeneous equation is

$$p(t) = A\, p(t-1).$$

By using a two-tuple vector c of constants to be determined by the initial conditions, we can solve the homogeneous equation system by iteration as

t=0	p(0) = c
t=1	p(1) = A c
t=2	p(2) = AA c = A^2 c
t=s	p(s) = A^s c.

Thus for any t, $p(t) = A^t c$. The complete solution is

$$p(t) = A^t c + p^* = A^t c + (I-A)^{-1} b. \qquad (3.7)$$

If we let c be determined by the initial condition $p(0) = p_0$, then

$$p(0) = p_0 = c + (I-A)^{-1} b \qquad \text{(setting t=0 in (3.7))}$$

or,

$$c = p_0 - (I-A)^{-1} b.$$

Hence

$$p(t) = A^t [p_0 - (I-A)^{-1} b] + p^*.$$

Suppose each element of the matrix A^t goes to zero as t becomes infinity, then

$$\lim_{t \to \infty} p^*t) = p^* = (I-A)^{-1} b.$$

In other words the system of prices converges to the steady-state solution p^* i.e. the simultaneous demand-supply system is stable. The condition that each element of the matrix A^t goes to zero as $t \to \infty$, is equivalent to the condition that each characteristic root (or eigenvalue) of the coefficient matrix A is less than one in absolute value. This can be seen in another way. Try the solution of the linear system (3.5) in the form $p(t) = \lambda^t c$, where λ is a scalar to be determined and c is an n-tuple vector to be determined by the initial conditions for the complete solution. If this solves the system (3.5), then by substitution we must have for the homogeneous part

$$\lambda^t c - A \lambda^{t-1} c = (\lambda I-A) \lambda^{t-1} c = 0$$

which yields $(\lambda I-A) c = 0$, since λ^{t-1} cannot be zero for any finite t and any non-zero λ. Thus we obtain the characteristic equation $(\lambda I-A) c = 0$, which implies equivalently that the determinant of $(\lambda I-A)$ is zero. As an example let $A = \begin{bmatrix} 4 & -2 \\ 1 & 1 \end{bmatrix}$

then the determinantal equation is

$$0 = |\lambda I-A| = \begin{vmatrix} \lambda-4 & 2 \\ -1 & \lambda-1 \end{vmatrix} = \lambda^2 - 5\lambda + 6$$

which gives two roots $\lambda_1 = 2$, $\lambda_2 = 3$. For $\lambda_1 = 1$ we have the eigenvector $c(1) = \binom{1}{1}$, whereas for $\lambda_2 = 3$, the eigenvector is $c(2) = \binom{2}{1}$. The solution of the homogeneous equations can therefore be written as

$$p_H(t) = k_1 \binom{1}{1} 2^t + k_2 \binom{2}{1} 3^t$$

where k_1 and k_2 are independent constants to be determined by the initial conditions. Let the steady-state solution vector be $p^* = \begin{pmatrix} p_1^* \\ p_2^* \end{pmatrix}$. Then the complete solution is

$$\begin{pmatrix} p_1(t) \\ p_2(t) \end{pmatrix} = k_1 \binom{1}{1} 2^t + k_2 \binom{2}{1} 3^t + \begin{pmatrix} p_1^* \\ p_2^* \end{pmatrix}$$

where k_1 and k_2 are to be determined by the two initial values say $p_1(0) = p_{10}$, $p_2(0) = p_{20}$, i.e.

$$k_1 = -(p_{10} - p_1^*) + 2(p_{20} - p_2^*)$$
$$k_2 = (p_{10} - p_1^*) - (p_{20} - p_2^*).$$

3.2 Linear Differential Equation Systems

Linear differential equations are very similar to the difference equations except for two things. One is that $Ae^{\lambda t}$ replaces $A\lambda^t$ in the trial solution, where e = 2.718 is the base of natural or Naperian logarithm. The second is that the particular solution can be found in many cases by direct integration when the nonhomogeneous term is a continuous function of time. For the first order linear differential equation with a constant coefficient and a constant nonhomogeneous term

$$(D - a) y(t) = b \qquad D = d/dt \tag{4.1}$$

the homogeneous part $(D-a) y=0$ has the solution $y = y(t) = Ae^{at}$, which may also be found out by integrating both sides of the equation

$$\frac{dy}{dt} = ay \tag{4.2}$$

with A being a constant of integration. Alternatively, we may try the trial solution $y_t = Ae^{\lambda t}$ and substitute in the homogeneous equation (4.2) to get

$$Ae^{\lambda t} (\lambda - a) = 0$$

yielding $\lambda = a$. The homogeneous part has then the solution $y_H = Ae^{at}$. Next we consider the particular solution y_p. If a steady-state level y^* exists at which Dy^*

= 0, we have $y_p = y^*$. From (4.1) one obtains $y^* = -b/a$, provided a is not zero. The complete solution is therefore

$$y_t = y_H + y_p = Ae^{at} - b/a \qquad (4.3)$$

where A is to be determined by one initial condition. If it is determined by y_o then it follows from (4.3) by setting t=0 that $A = (y_o + b/a)$. If the coefficient a is zero, then the particular solution cannot be determined by this way. Either we integrate the equation $\frac{dy}{dt} = b$ (since a=0) to obtain $y(t) = bt + c$, where c is an arbitrary constant to be determined, or we try the time-dependent solution $y_p = kt$ in (4.1) as in the case of a difference equation to obtain

$$k - akt = k(1-at) = b$$

but since a=0, k=b. With a=0, the homogeneous solution is $y_H = Ae^{0t} = A$. The complete solution is $y_t = A + bt$, or on using the initial value y_o we get the complete solution

$$y(t) = y_o + bt.$$

3.2.1 Second Order Equation

The second order linear differential equation

$$(D^2 + a_1 D + a_2) y(t) = b \qquad (4.4)$$

with constant coefficients a_1, a_2 and a constant homogeneous term b can be solved by attempting the trial solution $y(t) = Ae^{\lambda t}$, where A is a constant to be determined by the initial conditions. On using the trial solution in the homogeneous part of equation (4.4) one gets $(\lambda^2 + a_1 \lambda + a_2) Ae^{\lambda t} = 0$, which yields the quadratic characteristic equation in λ

$$\lambda^2 + a_1 \lambda + a_2 = 0. \qquad (4.5)$$

This has two roots $\lambda_1, \lambda_2 = \frac{-a_1 \pm \sqrt{a_1^2 - 4a_2}}{2}$. Assuming the roots to be real and distinct, we may obtain the solution of the homogeneous equation as a linear combination of $A_1 e^{\lambda_1 t}$ and $A_2 e^{\lambda_2 t}$

$$y_H = A_1 e^{\lambda_1 t} + A_2 e^{\lambda_2 t}$$

where A_1, A_2 are two independent constants to be determined by the initial conditions. If a_2 is not zero, then the steady-state equilibrium y^* exists at which $y_p = y^*$ and $Dy^* = 0 = D^2 y^*$. From (4.4), $y^* = b/a_2$. Hence the general solution is $y(t) = y_H + y^*$ or,

$$y(t) = A_1 e^{\lambda_1 t} + A_2 e^{\lambda_2 t} + b/a_2. \tag{4.6}$$

If two initial conditions y_0 and $y_0' = Dy$ at $t=0$ are given, then A_1 and A_2 can be found by solving the two linear equations

$$y_0 = A_1 + A_2 + b/a_2$$
$$y_0' = \lambda_1 A_1 + \lambda_2 A_2 + b/a_2$$

i.e.

$$A_1 = (\lambda_2 - \lambda_1)^{-1} \{\lambda_2 (y_0 - b/a_2) - (y_0' - b/a_2)\}$$
$$A_2 = (\lambda_2 - \lambda_1)^{-1} \{(y_0' - b/a_2) - \lambda_1 (y_0 - b/a_2)\}.$$

In the general situation the characteristic equation (4.5) has four cases depending on the discriminant $\Delta = a_1^2 - 4a_2$.

<u>Case 1</u> (distinct real roots): Here $\lambda_1 \neq \lambda_2$ and these roots are real. The complete solution is given by (4.6). It is clear that if each root is negative, then $e^{\lambda_i t}$ also denoted as $\exp(\lambda_i t)$ goes to zero as t tends to infinity and hence $y(t)$ converges to the steady-state equilibrium $y^* = b/a_2$. The path of $y(t)$ is then said to be stable (i.e. exponentially stable).

<u>Case 2</u> (equal real roots): Here $\lambda_1 = \lambda_2$ and it is also called repeated or double root. The homogeneous solution written as $y_H = A_1 e^{\lambda_1 t} + A_2 e^{\lambda_2 t} = (A_1 + A_2) e^{\lambda_1 t} = A_3 e^{\lambda_1 t}$ leaves us with only one constant $A_3 = A_1 + A_2$ and not two. But for a linear second order system we need two independent constants. Hence we try by analogy with the similar case of a difference equation

$$y_H = A_3 \exp(\lambda_1 t) + A_4 \, t \exp(\lambda_1 t)$$

where A_3, A_4 are two independent constants to be determined by the initial conditions. For example, the second order equation

$(D^2 - 8D + 16) y(t) = 0$

has the double root $\lambda_1 = 4$. Hence its solution is

$$y_H = (A_3 + A_4 t) \exp(4t). \qquad (4.7)$$

If the two initial conditions are given as $y_0 = 1$, $y'_0 = 6$, then we obtain from (4.7): $A_3 = y_0 = 1$ and $A_4 = y'_0 - 4A_3 = 6 - 4 = 2$ and hence the solution

$$y_t = y_H = (1 + 2t) e^{4t}.$$

Clearly the time-path is explosive since y_t goes to infinity as t gets infinitely large.

Case 3 (real roots of opposite sign): Here $\lambda_1 > 0$ and $\lambda_2 < 0$, so that the complete solution with a constant nonhomogeneous term appears as

$$y(t) = A_1 \exp(\lambda_1 t) + A_2 \exp(\lambda_2 t) + y^*.$$

If A_1, A_2 are positive constants, then the second component $\exp(\lambda_2 t)$ goes to zero, but the first component $\exp(\lambda_1 t)$ explodes to infinity as t goes to infinity. This is then called a __saddle point__. In optimal control models of the type known as LQG (linear quadratic Gaussian) this sort of solution characterizes the optimal path but in that case a separate condition known as the transversality condition implies that A_1 is zero, so that the optimal path is stable.

Case 4 (complex roots): The two roots λ_1, λ_2 in this case are complex conjugate i.e. they occur as solutions of the characteristic equation (4.5) with $a_1^2 < 4a_2$. The roots then become

$$\lambda_1 = \alpha + \beta i, \ \lambda_2 = \alpha - \beta i, \ i = +\sqrt{-1}$$

where α, β are real numbers such that $\alpha = -a_1/2$ and $\beta = (½)\sqrt{4a_2 - a_1^2}$. Since one can write $\exp(\pm \beta i t)$ as $(\cos\beta t \pm i \sin\beta t)$, therefore the homogeneous solution can be expressed as

$$\begin{aligned} y_H &= A_1 \exp((\alpha + \beta i)t) + A_2 \exp((\alpha - \beta i)t) \\ &= \exp(\alpha t) [A_1 (\cos\beta t + i \sin\beta t) + A_2 (\cos\beta t - i \sin\beta t)] \\ &= \exp(\alpha t) [A_3 \cos\beta t + A_4 \sin\beta t] \end{aligned} \qquad (4.8)$$

where $A_3 = A_1 + A_2$ and $A_4 = (A_1 - A_2)i$ are the two constants to be determined by the initial conditions. Consider an example

$(D^2 - 4D + 13) y(t) = 26$, $y_o = 1$, $y'_o = 8$.

The characteristic equation $\lambda^2 - 4\lambda + 13 = 0$ has two roots $2 \pm 3i$. Combining the homogeneous solution $y_H = \exp(2t)[A_3 \cos 3t + A_4 \sin 3t]$ and the particular solution $y_p = y^* = 26/13 = 2$ we obtain the general solution

$$y_t = \exp(2t) [A_3 \cos 3t + A_4 \sin 3t] + 2. \qquad (4.9)$$

To determine the constants A_3, A_4 we use the two given initial conditions $y_o = y(t)$ at $t = 0$ and $y'_o = Dy_t$ at $t = 0$, where $D = d/dt$. First, by setting $t = 0$ in the general solution (4.8) we get

$$y_o = y(0) = e^o [A_3 \cos 0 + A_4 \sin 0] + 2$$
$$= A_3 + 2, \text{ since } \cos 0 = 1 = e^o, \sin 0 = 0$$

since $y_o = 1$, $A_3 = -1$. Next we differentiate (4.8) and then set $t=0$:

$$y'(t) = Dy(t) = 2 \exp(2t)[A_3 \cos 3t + A_4 \sin 3t]$$
$$+ \exp(2t) [A_3 (-3 \sin 3t) + 3A_4 \cos 3t]$$

since $D(\cos 3t) = -3 \sin 3t$, $D(\sin 3t) = 3 \cos 3t$ hence at $t=0$, $y'_o = 2A_3 + 3A_4 = -2 + 3A_4$ but $y'_o = 8$ hence $A_4 = 10/3$. The final general solution is therefore

$$y(t) = \exp(2t)[-\cos 3t + (10/3) \sin 3t] + 2.$$

Some comments on the general form of the solution (4.8) are in order. First, we note that the term $\cos \beta t$ is a circular function of (βt) with period 2π in radian measure and amplitude 1. The period of 2π means that the graph will repeat itself in shape after 2π periods i.e. $\cos(\theta + 2n\pi) = \cos\theta$ and also $\sin(\theta + 2n\pi) = \sin\theta$ for any integer n and any angle θ. Any complex number $\lambda = \alpha \pm \beta i$ can be written in any of the two forms

$$\lambda = \alpha \pm \beta i = R(\cos\theta \pm i \sin\theta) = R \exp(\pm i\theta)$$

where $R = |\lambda| = \sqrt{(\alpha^2 + \beta^2)}$ and θ is an angle measured in radians. The angle θ is called the amplitude of the complex number and can be found from $\cos\theta = \beta/R = \beta/\sqrt{(\alpha^2 + \beta^2)}$. Thus $\cos(2n\pi) = 1$ and $\sin(2n\pi) = 0$ for all integer n. Now when $\cos(\beta t)$ is multiplied by the constant A_3 on the right-hand side of (4.8), it causes the amplitude to change from ± 1 to $\pm A_3$. Since 2π is the common period for both sine and cosine curves to repeat, the term $[A_3 \cos\beta t + A_4 \sin\beta t]$ will also show a repeating or periodic cycle every 2π periods. Secondly, the question whether the overall path of y_t in (4.8) (or (4.9)) converges or not depends on the index α of the first term $\exp(\alpha t)$ (or, $\exp(2t)$). If $\alpha > 0$, then the term $\exp(\alpha t)$ explodes to infinity as t gets infinite. This will produce a magnifying effect on the amplitude of $(A_3 \cos\beta t + A_4 \sin\beta t)$ and cause deviations from the equilibrium level $y^* = y_p$ to increase in each successive cycle. If $\alpha = 0$, the fluctuation will be uniform and of a constant amplitude ±1. But if $\alpha < 0$, then the path will follow a damped fluctuation finally converging to the steady-state equilibrium level y^*. Thus the condition of stability (or convergence) is that the real part i.e. α in $\alpha \pm \beta i$ of every root must be negative. This covers the case of real roots also since $\beta = 0$ for the real root. Clearly, the solution in (4.9) is unstable i.e. explosive.

As in the case of linear difference equations we need some care in deriving a particular solution when the nonhomogeneous term is a given time function b(t) in (4.4). We again follow the method of undetermined coefficients which we illustrate by two examples. To find a particular solution y_p of the second-order equation

$$(D^2 + 4D - 1) y(t) = t^2 - 3t + 4$$

we note that the nonhomogeneous term b(t) is a polynomial i.e. a quadratic function of t. Hence we assume a particular solution of the form

$$y_p = At^2 + Bt + C$$

and seek to find the undetermined constants A, B and C. Differentiating gives us

$$y_p' = Dy_p = 2At + B, \quad y_p'' = D^2 y_p = 2A$$

and substituting into the given differential equation produces

$$2A + 8At + 4B - At^2 - Bt - C = t^2 - 3t + 4.$$

Next we equate coefficients of like powers of t. The justification for this is that the functions 1, t and t^2 are linearly independent; hence the equation $c_1 + c_2 t + c_3 t^2 = 0$ can be satisfied only for $c_1 = c_2 = c_3 = 0$. In the present case $(2A + 4B - C - 4) + (8A - B + 3)t + (-A-1)t^2 = 0$ which yields $A = -1$, $B = -5$, $C = -26$. Hence the particular solution is

$$y_p = y_p(t) = -t^2 - 5t - 26.$$

As a second example consider the equation

$$(D^2 + 3D - 4) y(t) = 3 \exp(t)$$

the homogeneous solution is

$$y_H(t) = c_1 \exp(t) + c_2 \exp(-4t)$$

where c_1, c_2 are constants to be determined by two initial conditions. Note that we cannot choose for a particular solution a constant times $\exp(t)$ since we already have such a term present in $y_H(t)$. Instead we adopt the trick applied before and set $y_p(t) = At \exp(t)$ where A is yet to be determined. When used in the complete original equation above we get

$$(At + 2A + 3At + 3A - 4At) \exp(t) = 3 \exp(t)$$

yielding $A = 3/5$. Hence the general solution is

$$y(t) = c_1 \exp(t) + c_2 \exp(-4t) + (3/5)t \exp(t).$$

3.2.2 n-th Order Equation

The case of an n-th order linear differential equation with constant coefficients

$$(D^n + a_1 D^{n-1} + \ldots + a_{n-1} D + a_n) y(t) = b \qquad (5.1)$$

can be solved by methods developed earlier. To illustrate we assume for simplicity that b is a constant and the steady state solution $y^* = b/a_n$ exists. The n roots λ_i ($i=1,2,\ldots,n$) of the n-th order characteristic equation

$$\lambda^n + a_1 \lambda^{n-1} + \ldots + a_{n-1} \lambda + a_n = 0 \qquad (5.2)$$

have to be used in the solution y_H of the homogeneous form with $b = 0$. If all roots are real and distinct, then

$$y_H = \sum_{i=1}^{n} A_i \exp(\lambda_i t) \qquad (5.3)$$

where the n constants A_i ($i=1,2,\ldots,n$) are to be determined by n initial values e.g. $y_0, Dy_0, \ldots, D^{n-1}y_0$.

But as before some modifications of (5.3) are needed when some of the n roots are not all real and distinct. First, suppose there are four repeated roots say $\lambda_1 = \lambda_2 = \lambda_3 = \lambda_4$. Then we must have the first four terms of y_H in (5.3) as $[A_1 \exp(\lambda_1 t) + A_2 t \exp(\lambda_1 t) + A_3 t^2 \exp(\lambda_1 t) + A_4 t^3 \exp(\lambda_1 t)]$. In general, if m roots

($m < n$) are equal, then the first m terms on the right-hand side of (5.3) would be $\exp(\lambda_1 t)[A_1 + A_2 t + A_3 t^2 + \ldots + A_m t^{m-1}]$. Second, suppose that the first two roots λ_1, λ_2 are complex conjugate pairs $\alpha \pm \beta i$, α and β being real numbers. Then the first two terms on the right-hand side would be $\exp(\alpha t)[A_1 \cos\beta t + A_2 \sin\beta t]$.

The condition of stability remains the same. If any root, real or complex is written as $\lambda_j = \alpha_j + \beta_j i$ where α_j, β_j are real numbers (j=1,2,...,n) such that $\beta_j = 0$ for λ_j real and β_j is negative or positive for λ_j complex, then the condition for stability is that $\alpha_j < 1.0$ i.e. the real part of every root must be negative. But this method of stability analysis requires us to compute all the n characteristic roots. A simpler method is provided by the <u>Routh-Hurwicz Theorem</u> which says that the real parts of all the roots or solution of the n-th degree polynomial equation

$$a_0 \lambda^n + a_1 \lambda^{n-1} + \ldots + a_{n-1} \lambda + a_n = 0$$

are negative if and only if (i.e. the condition is both necessary and sufficient) the first n of the following sequence of determinants

$$|a_1| \quad \begin{vmatrix} a_1 & a_3 \\ a_0 & a_2 \end{vmatrix}, \quad \begin{vmatrix} a_1 & a_3 & a_5 \\ a_0 & a_2 & a_4 \\ 0 & a_1 & a_3 \end{vmatrix}, \quad \begin{vmatrix} a_1 & a_3 & a_5 & a_7 \\ a_0 & a_2 & a_4 & a_6 \\ 0 & a_1 & a_3 & a_5 \\ 0 & a_0 & a_2 & a_4 \end{vmatrix}$$

are all positive. Thus for the second order equation $(D^2 - 4D - 12) y_t = 0$, the second-degree (n=2) characteristic equation is $\lambda^2 - 4\lambda - 12 = 0$ and $a_0 = 1$, $a_1 = -4$, $a_2 = -12$. We have to consider therefore the first two determinants a_1 = determinant of $a_1 = a_1 = -4$

$$\begin{vmatrix} a_1 & a_3 \\ a_0 & a_2 \end{vmatrix} = \begin{vmatrix} -4 & 0 \\ 1 & -12 \end{vmatrix} = 48 \quad \text{(note } a_3 = 0\text{).}$$

Since the first determinant is not positive, the system is not stable. Note that $|a_1|$ denotes here not the absolute value of a_1 but the determinant of a_1 which is a_1 itself. Further we should take $a_m = 0$ for all $m > n$ e.g. in the above case $a_3 = 0$.

3.2.3 Simultaneous Equations

As in simultaneous difference equations, the simultaneous linear differential equations can be solved in two ways. One is to reduce the n first order linear differential equations to a single n-th order equation and apply the method in the earlier section. This is always possible by the superposition theorem. The second method is to apply the matrix approach. Let us solve the homogeneous equation in 3-tuple vector $y = y(t)$:

$$Dy(t) = Ay, \quad A = \begin{bmatrix} -4 & 5 & 5 \\ -5 & 6 & 5 \\ -5 & 5 & 6 \end{bmatrix}. \tag{5.4}$$

Assume a solution of the form $y(t) = c \exp(\lambda t)$, where λ is a scalar, $\exp(\lambda t) = e^{\lambda t}$ and $c^T = (c_1, c_2, c_3)$ is a 3-tuple row vector (T denoting transpose). On substitution of y_t and $D y(t) = \lambda c \exp(\lambda t)$ in (5.4) we get $(\lambda I - A) c \exp(\lambda t) = 0$. This yields the characteristic equation $(\lambda I - A) c = 0$, which implies that λ must be an eigenvalue of matrix A. The third order square matrix A above must have three eigenvalues $\lambda_1 = 1$, $\lambda_2 = 1$, $\lambda_3 = 6$. Let $c(i)$ be the eigenvector of λ_i (i=1,2,3), then $c^T(1) = (1,0,1)^T$, $c^T(2) = (1,1,0)^T$ and $c^T(3) = (1,1,1)^T$. The complete solution which consists only of the homogeneous solution can therefore be written as

$$y = \begin{pmatrix} y_1 \\ y_2 \\ y_3 \end{pmatrix} = k_1 \begin{pmatrix} 1 \\ 0 \\ 1 \end{pmatrix} \exp(t) + k_2 \begin{pmatrix} 1 \\ 1 \\ 0 \end{pmatrix} \exp(t) + k_3 \begin{pmatrix} 1 \\ 1 \\ 1 \end{pmatrix} \exp(6t)$$

where k_1, k_2, k_3 are to be determined by the three initial values of $y_1(0)$, $y_2(0)$, $y_3(0)$ say.

Sometimes it is possible to diagonalize the square matrix A by a transformation $y = Px$, where P is a nonsingular square matrix such that $P^{-1} AP$ equals a diagonal matrix G say with diagonal elements g_i say and then the problem reduces to solving each linear differential equation separately. For the example (5.4), $y = Px$ implies $Dy = PDx$ and hence (5.4) reduces to $Dx = P^{-1} AP\, x_t = G\, x_t$. By using the transformation matrix

$$P = \begin{bmatrix} -4 & -3 & 0 \\ 5 & 4 & 1 \\ 7 & 2 & 1 \end{bmatrix}$$

we get

$$G = P^{-1} AP = \begin{bmatrix} -2 & 0 & 0 \\ 0 & -1 & 0 \\ 0 & 0 & 2 \end{bmatrix}$$

i.e. $g_1 = -2$, $g_2 = -1$, $g_3 = 2$.

Hence the reduced form linear equation system is

$$\frac{dx_1}{dt} = -2x_1, \quad \frac{dx_2}{dt} = -x_2, \quad \frac{dx_3}{dt} = 2x_3$$

with the solution vector $x^T = (k_1 e^{-2t}, k_2 e^{-t}, k_3 e^{2t})^T$ where k_1, k_2, k_3 are to be determined by the initial conditions. Finally, by using the transformation $y = Px$ we get back the general solution

$$y = Px = k_1 \begin{pmatrix} -4 \\ 5 \\ 7 \end{pmatrix} \exp(-2t) + k_2 \begin{pmatrix} -3 \\ 4 \\ 2 \end{pmatrix} \exp(-t) + k_3 \begin{pmatrix} 1 \\ 0 \\ -1 \end{pmatrix} \exp(2t).$$

If the initial condition $y_0^T = (2, -12, 24)^T$ is given for determining k_1, k_2, k_3 then the completely definite general solution can be written as

$$y = 4 \begin{pmatrix} -4 \\ 5 \\ 7 \end{pmatrix} e^{-2t} - 6 \begin{pmatrix} -3 \\ 4 \\ 2 \end{pmatrix} e^{-t} - 8 \begin{pmatrix} 0 \\ 1 \\ -1 \end{pmatrix} e^{2t}.$$

Next consider the general case of the n linear first-order equations in the n-tuple vector y with a constant nonhomogeneous n-tuple vector b such that the steady-state equilibrium solution y^* exists as a particular solution

$$Dy = Ay - b.$$

Try the solution $y = c \exp(\lambda t)$ for the homogeneous part $Dy = Ay$ to obtain the characteristic equation $(\lambda I - A) c = 0$ which must have n characteristic roots or eigenvalues λ_i (i=1,2,...,n), for each of which we have a characteristic vector or eigenvector $c(i)$. If the eigenvalues are all distinct, then the homogeneous solution vector y_H can be written as

$$y_H = \sum_{i=1}^{n} k_i c(i) \exp(\lambda_i t)$$

where k_i (i=1,2,...,n) are the n independent constants to be determined by the initial value $y_0 = y(0)$ of the n-tuple vector y at $t = 0$. The particular solution is $y^* = y_p = A^{-1} b$ assuming A to be nonsingular. Hence the complete solution is

$$y = y_H + y_p = \sum_{i=1}^{n} k_i c(i) e^{\lambda_i t} + y^*.$$

Given the initial vector $y_0 = y(t)$ at $t = 0$ we can solve for the n values $k_1, k_2,...,k_n$ from the equation

$$y_0 = \sum_{i=1}^{n} k_i \, c(i) + y^*.$$

Two applications of the simultaneous linear differential equations are most common in economics. One is the Walrasian price adjustment model in n competitive markets with linear demand d and supply s vectors

$$d = a - Ap; \quad s = -b + Bp$$

where A and B are square matrices and it is assumed that excess demand (supply) tends to increase (decrease) the price vector

$$Dp = K(d - s) = K(a + b) - K(A + B) \, p$$

where K is a diagonal matrix with positive diagonal elements k_i denoting the speed of adjustment. The second application is in the dynamic version of the Leontief input-output model. Recall that in Chapter One we had the static version of Leontief's input-output (IO) model

$$x_i = \sum_{j=1}^{n} x_{ij} + d_i = \sum_{j=1}^{n} a_{ij} \, x_j + d_i; \quad i=1,2,\ldots,n$$

where the gross output (x_i) of sector i is used partly as raw material demand by all the other sectors including sector i ($= \sum_{j=1}^{n} a_{ij} \, x_j$) and partly as final demand d_i representing household consumption, investment and net exports over imports. Here $a_{ij} x_j$ represents that part of output of sector i which is demanded by sector j as inputs or raw materials. But this demand is for current inputs only. If a part of output x_i is used for capital inputs rather than current inputs, then it should be related to the <u>increase</u> of outputs in sector j; thus $b_{ij} \, \Delta x_j$ is the incremental demand of input i required by the output increase (Δx_j) planned by sector j, where b_{ij} is the incremental capital-output ratio (i.e. it is the capital-account equivalent of the current-account IO coefficient a_{ij}). On using a subscript t for time and approximating Δx_j by the time-derivative $dx_j/dt = Dx_j(t)$ one can express the dynamic IO equations as

$$x_i(t) = \sum_{j=1}^{n} a_{ij} \, x_j(t) + \sum_{j=1}^{n} b_{ij} \, Dx_j(t) + d_i.$$

DYNAMIC SYSTEMS IN ECONOMICS   115

In vector-matrix notation

$$x(t) = A\,x(t) + B\,D\,x(t) + d; \qquad A = (a_{ij}),\ B = (b_{ij})$$

or, $\quad D\,x(t) = B^{-1}((I-A)\,x(t) - d).$

This is a simultaneous system of n linear difference equations, for which the stability could be discussed in terms of the characteristic roots of the matrix $B^{-1}(I-A)$, where it is assumed that the capital coefficient matrix B is nonsingular.

An alternative formulation is to use a discrete time version $(x_j(t+1) - x_j(t))$ of $\Delta x_j(t)$ and obtain a simultaneous system of n first order difference equations as

$$x(t) = A\,x(t) + B[x(t+1) - x(t)] + d(t)$$

where x(t) is an n-tuple vector and d(t) is a given set of demands which may be time dependent. The homogeneous equation system is

$$x(t+1) = (I+C)\,x(t), \qquad C = B^{-1}(I-A)$$

which has the solution

$$x(t) = (I+C)^t\,v$$

where the vector v is determined by the initial conditions. For stability each characteristic root of (I+C) must be less than one in absolute value. If the demand vector d(t) grows exponentially i.e. $d_i(t+1) = (1 + r_i)\,d_i(t)$, or $d(t+1) = (I+R)\,d(t)$, R being a diagonal matrix with positive diagonal elements, then a particular solution $x_p(t)$ can be derived as

$$x_p(t) = (I+C)^t\,x(0) - \sum_{s=0}^{t-1} (I+C)^s\,B^{-1}\,d(t-1-s)$$

by an iterative method. Another particular solution derived by Richard Stone is

$$x_p = (I-A)^{-1}\,[\sum_{s=0}^{\infty} (B\,(I-A)^{-1})^s\,R^s]\,d$$

This has an interesting economic interpretation. To supply the final demand vector d, a current output of $(I-A)^{-1}\,d$ is required. In the next and all subsequent periods there will be increased final demand equal to Rd. This will require an increase in output of $(I-A)^{-1}\,Rd$ next period, implying an increase in investment of $B(I-A)^{-1}\,Rd$ this period and so on. Various properties of the complete solution may thus be analyzed.

Three general comments may be made about the stability and fluctuations implied by the solutions of the linear difference and differential equations with constant coefficients and a constant steady state equilibrium. First of all, if the characteristic roots are complex conjugate i.e. $\alpha \pm \beta i$, $i = +\sqrt{-1}$, then for the differential case the real part α should be negative for stability for each characteristic root; in the difference equation case we require $R = \sqrt{(\alpha^2 + \beta^2)}$ to be less than one (note that R is also called the absolute value of the complex roots $\alpha \pm \beta i$) for each characteristic root. Secondly, a first order linear difference equation $Dy(t) - ay(t) = 0$ can generate oscillations in the solution path $y(t) = y(0) \, a^t$ if $a < 0$ in the sense of positive and negative values, but a first order linear differential equation can never do that since its solution is $y_t = y_o \exp(at)$ which is exponential (rising or decaying steadily if a is nonzero). Thirdly, it is sometimes possible to use the differential approximation dy/dt for the finite difference $[y(t+1) - y(t)]$ and solve difference equations, particularly the non-linear ones in terms of known nonlinear differential equations.

3.3 Economic Applications

The economic applications of linear difference and differential equations are intended to illustrate the following aspects of economic behavior: (a) role of expectations in the sense of adaptive and rational expectations, (b) lags in adjustment and self-correction through learning, (c) various types of adaptive and cautious control designed to influence the output or state behavior and (d) different types of stability of an economic system. We first discuss some examples in microeconomic theory and then in the macro-dynamic economic systems.

3.3.1 Market Model with Price Expectations

Consider the market model for a good where supply s(t) responds to expected $\hat{p}(t)$ rather than current price p(t) and demand d(t) is linear in current price

$$d(t) = a_1 - b_1 \, p(t), \qquad s(t) = -a_2 + b_2 \, \hat{p}(t)$$
$$d(t) = s(t) \tag{6.1}$$

where all the coefficients a_i, b_i are assumed positive. Note that if $\hat{p}(t) = p(t-1)$ i.e. price expected at time t is the last period's realized price, then we get back the cobweb model as before. But the assumption $\hat{p}(t) = p(t-1)$ for every t seems very implausible as a hypothesis of expectation formation. Two more realistic hypotheses have been proposed for determining the expected price $\hat{p}(t)$. One is the model of **adaptive expectations** due to Nerlove

$$\hat{p}(t) = \hat{p}(t-1) + c(p(t-1) - \hat{p}(t-1))$$
$$= c\, p(t-1) + (1-c)\, \hat{p}(t-1), \qquad 0 \le c \le 1 \qquad (6.2)$$

where c is a positive coefficient not greater than one. This equation says that the expected price in each period is revised on the basis of the error between the observed value and the previously expected value. Thus $\hat{p}(t)$ is a linearly weighted average of the two prices $p(t-1)$ and $\hat{p}(t-1)$. If $p(t-1)$ is greater (less) than the expected price $\hat{p}(t-1)$ in period t-1, then the new expected value is revised upwards (downwards).

A second hypothesis is that the expected price $\hat{p}(t)$ follows a second order process as follows

$$\hat{p}(t) = p(t-1) + k(p(t-1) - p(t-2)) \qquad (6.3)$$

where the constant k, which can take any value positive, negative or zero is sometimes called the coefficient of expectation. This hypothesis does not have the restriction that c had in (6.2) i.e. $0 \le c \le 1$.

Considering the first hypothesis, we derive from the supply equations $s(t) = -a_2 + b_2 \hat{p}(t)$, and $s(t-1) = -a_2 + b_2 \hat{p}(t-1)$ the values of $\hat{p}(t) = (s(t) + a_2)/b_2$ and $\hat{p}(t-1) = (s(t-1) + a_2)/b_2$ and substituting these in (6.2) we get

$$s(t) = (1-c)\, s(t-1) - a_2 c + b_2 c\, p(t-1) \qquad (6.4)$$

but $s(t) = d(t) = a_1 - b_1 p(t)$ for all t, hence $s(t-1) = a_1 - b_1 p(t-1)$. On using these in (6.4) we obtain the final reduced from equation in p(t)

$$p(t) - h_1 p(t-1) = h_2 \qquad (6.5)$$

where

$$h_1 = 1 - c\left(1 + \frac{b_2}{b_1}\right), \quad h_2 = \frac{c(a_1 + a_2)}{b_1}.$$

The general solution is

$$p(t) = (p_0 - p^*)\, h_1^t + p^*$$

where p^* is the steady-state particular solution $p^* = h_2/(1-h_1)$, assuming $h_1 \ne 1$ and p_0 is the initial value. The stability condition is $h_1 < 1$ i.e.

$$-1 < 1 - c(1 + b_2/b_1) < +1$$

After some algebraic manipulation this can be reduced to

$$1 - \frac{2}{c} < -\frac{b_2}{b_1} < 1. \tag{6.6}$$

If we used the assumption of the original cobweb model i.e. $\hat{p}(t) = p(t-1)$, all t, then the stability condition would have been

$$-1 < -b_2/b_1 < 1. \tag{6.7}$$

But since $0 < c < 1$, we have $2/c > c$ and so $1 - (2/c) < -1$. This shows that the stability condition (6.6) is less stringent than (6.7) as far as the left side is concerned and it is the left side which matters most in case of most empirical demand and supply functions. Hence we conclude that the adaptive expectations hypothesis makes the model more stable and hence speeds up the process of convergence of $p(t)$ to the steady-state equilibrium level p^*.

In case of the second type of expected price formation equation (6.3), the final reduced form equation is of second order

$$-b_1 p(t) - b_2(1+k) p(t-1) + b_2 k p(t-2) = -(a_1 + a_2).$$

The particular solution is $p^* = (a_1 + a_2)/(b_1 + b_2)$. The characteristic equation of the corresponding homogeneous equation is

$$\lambda^2 + \frac{b_2(1+k)}{b_1} \lambda - \frac{b_2 k}{b_1} = 0. \tag{6.8}$$

By using Schur Theorem one obtains the following inequalities for stability

$$1 + \frac{b_2}{b_1} > 0, \qquad 1 + \frac{b_2 k}{b_1} > 0, \qquad 1 - \frac{b_2(1+2k)}{b_1} > 0. \tag{6.9}$$

Two cases may be clearly distinguished according as $k > 0$ and $k < 0$. For $k > 0$, the first two inequalities of (6.9) are necessarily satisfied. By noting that the discriminant Δ of the quadratic equation (6.8) is positive for a positive k, where

$$\Delta = \frac{b_2^2 (1 + k)^2}{b_1^2} + \frac{4 b_2 k}{b_1}$$

the roots are real and distinct and since the succession of the signs of the coefficients in (6.8) is + + -, one root is negative and the other positive. We have a saddle point.

If $k < 0$, the first inequality in (6.9) is always satisfied since it does not involve k. Hence by applying the second and third inequality one can derive the stability conditions as

$$\text{for } k \leq -1/3, \quad b_2/b_1 < -1/k$$
$$-1/3 \leq k < 0, \quad b_2/b_1 < 1/(1+2k).$$

3.3.2 Cournot Game with Learning

As a solution of the imperfectly competitive model known as duopoly (two players) or oligopoly (many players), Augustin Cournot, an early nineteenth-century French economist offered a zero-sum game formulation. To take the simplest case of duopoly, two firms (i.e. two players) having identical total cost $TC(x_i) = c_0 + cx_i$, c_0, $c > 0$ face the total market demand

$$p = a - b(x_1 + x_2) \qquad (7.1)$$

where p is the price and $(x_1 + x_2)$ is the total supply by the two firms. Profit is $z_i = px_i - (c_0 + cx_i)$ for i=1,2. Player one maximizes his own profit π_1, conditional on a conjectural guess of his rival's output $\hat{x}_2$ say. Likewise, player two maximizes his profit π_2 conditional on his guess $\hat{x}_1$ of his rival's output. Besides this assumption of conjectural guesses, Cournot assumes an iterative procedure for revising the strategies by each player as follows; so long as the guesses are not equal to the actual or realized values (i.e. $\hat{x}_i \neq x_i$), the players go on revising their own strategies and the condition $\hat{x}_i = x_i$ for i=1,2 is used to define an equilibrium solution. Cournot showed that under normal shapes of demand functions and the rationality motive of profit maximization by each player, such an equilibrium point would exist even in the case where the number of players $n \geq 2$. For the two-player case, we obtain for each player the following necessary conditions for maximum profits

$$\frac{\partial \pi_1}{\partial x_1} = 0 \text{ i.e. } x_1 = (a_1/2b) - (1/2)\hat{x}_2; \quad a_1 = a - c > 0$$

$$\frac{\partial \pi_2}{\partial x_2} = 0 \text{ i.e. } x_2 = (a_1/2b) - (1/2)\hat{x}_1. \qquad (7.2)$$

On setting $\hat{x}_1 = x_1$, $\hat{x}_2 = x_2$ we obtain the equilibrium solution $x_1^* = x_2^* = a_1/3b$. The process of reaching this equilibrium solution by a sequence of trial and error simultaneous adjustments may be illustrated by a learning mechanism, assuming of course that the process converges to this equilibrium. Assume that each duopolist begins with an arbitrarily selected output level which he changes from one time point to another by looking at the gap between his desired and actual output level in the earlier time point

$$x_1(t) = x_1(t-1) + k[R_1(x_1(t-1)) - x_1(t-1)]$$
$$x_2(t) = x_2(t-1) + k[R_2(x_1(t-1)) - x_2(t-1)] \qquad (7.3)$$

where $R_1(x_2(t-1))$ is the reaction curve of player one given in (7.2) i.e. $x_1^d(t)$ = desired level = $(a_1/2b) - (1/2)\hat{x}_2(t-1)$ and $R_2(x_2(t-1))$ is $x_2^d(t) = (a_1/2b) - (1/2)\hat{x}_1(t-1)$. On using these values in (7.3), where for simplicity the same positive value of the adjustment coefficient is used, we obtain at equilibrium when $x_i(t) = \hat{x}_i(t)$, i=1,2

$$x_1(t) - (1-k) x_1(t-1) + 0.5k\, x_2(t-1) - \frac{ka_1}{2b} = 0$$

$$x_2(t) - (1-k) x_2(t-1) + 0.5k\, x_1(t-1) - \frac{ka_1}{2b} = 0.$$

In vector matrix terms

$$\begin{pmatrix} x_1(t) \\ x_2(t) \end{pmatrix} = \begin{bmatrix} 1-k & -k/2 \\ -k/2 & 1-k \end{bmatrix} \begin{pmatrix} x_1(t-1) \\ x_2(t-1) \end{pmatrix} + \begin{pmatrix} ka_1/2b \\ ka_1/2b \end{pmatrix}.$$

The characteristic equation is

$$\lambda^2 - 2(1-k)\lambda + (1-k)^2 - k^2/4 = 0$$

with two roots $\lambda_1 = 1-0.5k$, $\lambda_2 = 1-1.5k$. Hence the complete solution for one component $x_1(t)$ is

$$x_1(t) = c_1(1-0.5k)^t + c_2(1-1.5k)^t + a_1/3b \qquad (7.4)$$

where c_1, c_2 are to be determined by initial conditions. Also by symmetry $x_1(t) = x_2(t)$. It is clear from (7.4) that both roots are less than +1, since k is positive. Hence the system will be stable and converge to the Cournot equilibrium $x_1^* =$

$x_2^* = a_1/3b$, if the second root is greater than -1, which will be the case if k < 4/3. Otherwise it will diverge.

Two important objections to the Cournot model are that the parameters a,b in the demand function may not be completely known to the two players and also there may be additive error or noise such as

$$p = a - b(x_1 + x_2) + \text{error}$$

in (7.1), in which case the above condition (i.e. k < 4/3) may not suffice for convergence to the Cournot equilibrium. Moreover, the assumption of no retaliation by the rival seems very unrealistic for many practical situations. These and other imperfections have led to recent attempts at modifying and generalizing this simplified Cournot framework.

3.3.3 Model of Maximum Sustainable Yield

Like the Cournot model of competition and conflict, Volterra studied models in ecology with predator and prey. Such models yield growth paths known as logistic time paths. Recently, such models have been used in the optimal management of renewable resources like fisheries.

The predator-prey dynamics may be represented by the following pair of non-linear differential equations in the two populations, $x_1(t)$ for prey and $x_2(t)$ for the predator

$$\begin{aligned} Dx_1 &= r_1 x_1 (1 - x_1 - v x_2) - y_1 \\ Dx_2 &= r_2 x_2 (1 - x_2/x_1) - y_1 \end{aligned} \quad (8.1)$$

where $D = d/dt$ is the differential operator for $x_i = x_i(t)$, x_1 for the prey population (e.g., krill) and x_2 for the predator (e.g., whales) and y_i is the effective amount of fishing of the two species and r_i, v are positive parameters. We have to determine the maximum sustainable yield (MSY) at equilibrium when $Dx_1 = 0 = Dx_2$. Note the economic interpretation of the two nonlinear differential equations (8.1). First, the rate of change of the krill population (i.e. prey) is negatively related to both the size of the whale population (i.e. predator) and the level of fishing effort by fishermen. Secondly, the particular solution satisfies the equilibrium condition $Dx_1 = 0 = Dx_2$ and this yields

$$\begin{aligned} y_1 &= r_1 x_1 (1 - x_1 - v x_2) \\ y_2 &= r_2 x_2 (1 - (x_2/x_1)). \end{aligned} \quad (8.2)$$

Two types of policy rules may be analyzed on the basis of the particular solution (8.2). One is to determine the MSY for one population subject to a constant fishing effort. This leads e.g. to

$$\text{Max } y_2 = r_2 x_2 \left(1 - \frac{x_2}{x_1}\right)$$

$$\text{s.t. } 1 - x_1 - v x_2 = h_1, \ h_1 = \text{constant.} \tag{8.3}$$

A second situation is to maximize MSY for one population (e.g. the krill) for fixed yield of the other. This leads to the optimization model

$$\text{Max } y_1 = r_1 x_1 (1 - x_1 - v x_2)$$

$$\text{s.t. } y_2 = r_2 x_2 \left(1 - \frac{x_2}{x_1}\right) = h_2 \ (\text{constant}). \tag{8.4}$$

The first case of (8.3) may be solved by using the Lagrange multiplier λ for the constraint and applying the first order condition

$$\partial y_2/\partial x_i + \lambda(\partial h_1/\partial x_i) = 0 \qquad i=1,2.$$

This yields $r_2 x^2 - \lambda = 0$, $r_2 (1 - 2x) - v \lambda = 0$ where $x = x_2/x_1$. Eliminating λ one gets

$$v x^2 + 2x - 1 = 0 \ (\text{quadratic equation in } x).$$

Discarding a negative solution for x we end up with $x_1 = (1 + b) x_2$, $b = (1 + v)^{\frac{1}{2}}$. Thus we obtain the MSY levels at the equilibrium $Dx_1 = 0 = Dx_2$

$$y_2 = \frac{r_2 (1 + v - b)}{v} x_2$$
$$h_1 = 1 - (1 + v + b) x_2$$
$$y_1 = r_1 x_2 (1 + b) [1 - (1 + v + b) x_2].$$

Thus we obtain a parabolic relation between y_2 and the krill yield y_1 obtained under the constraint of constant effort.

In the second model i.e. (8.4) we follow the same method to obtain

$$r_1 (1 - 2x_1 - vx_2) + \lambda r_2 (x_2/x_1)^2 = 0$$

and

$$-r_1 v x_1 + \lambda r_2 \left(1 - 2\frac{x_2}{x_1}\right) = 0.$$

Eliminating λ we end up with a quadratic equation for x_1 which has the solution

$$x_1 = 0.25 [1 + (4 - v) x_2] \pm 0.25 [\{1 + (4 - v) x_2\}^2 - 8x_2 \{2 - 3v x_2\}]^{1/2}.$$

This example illustrates two important points. One is that the particular solution may be used as a static optimization problem. Second, the nonlinear differential equation is sometimes reducible to linear equation systems.

3.3.4 Keynesian Macro-dynamic Model

Two types of macro-dynamic Keynesian models have often been used. One is the **equilibrium** version, the other the **disequilibrium** model. In its simplest discrete time equilibrium version, illustrated by Samuelson's multiplier-accelerator interaction form the model has aggregate consumption ($C(t)$), investment ($I(t)$), national income ($Y(t)$) and autonomous government expenditure (G) as follows:

$$\begin{aligned} C(t) &= a\, Y(t-1) \qquad 0 < a < 1 \\ I(t) &= b\, (C(t) - C(t-1)) \\ Y(t) &= C(t) + I(t). \end{aligned} \qquad (9.1)$$

Here the positive parameters a and b denote the constant marginal propensity to consume and the acceleration coefficient respectively. All variables are measured in real terms. The first equation describes consumption behavior influenced by lagged income; the second equation indicates induced investment due to the increase in consumption demand and the third relation is an identity denoting demand-supply equilibrium. The reduced form equation is a second order linear difference equation

$$Y(t+2) - a(1 + b)\, Y(t+1) + ab\, Y(t) = G \qquad (9.2)$$

where G is assumed for simplicity to be a constant.

The particular solution Y^* is

$$Y^* = \frac{G}{1 - a(1 + b) + ab} = \frac{G}{1-a}$$

hence $\Delta Y^* = k\Delta G$, $k = (1-a)^{-1} > 1$ if $0 < a < 1$. Here k denotes the multiplier effect on income of a unit increase in autonomous government expenditure. According to Keynesian assumptions the multiplier effect on income is in real terms before the stage of full employment but it is operative in money terms only (through price rises only) after the stage of full employment. The homogeneous part of the equation (9.2) has the characteristic equation

$$\lambda^2 - a(1 + b)\lambda + ab = 0$$

with two roots λ_1, λ_2

$$\lambda_1, \lambda_2 = \frac{a(1 + b) \pm \sqrt{a^2(1+b)^2 - 4ab}}{2}$$

Several cases are now possible, as we discussed before

(i) Distinct real roots, when $a > 4ab/(1+b)^2$: nonoscillatory and nonfluctuating i.e. no periodic fluctuations.

(ii) Repeated real roots, when $a = 4b/(1+b)^2$: no oscillation and no periodic fluctuations.

(iii) Complex roots with damped (or increasing) fluctuations, when $a < 4b/(1+b)^2$ and $R = \sqrt{(ab)}$ is less than (or equal to) one and.

(iv) Complex roots with explosive fluctuations when $a < 4b/(1+b)^2$ and $R = \sqrt{(ab)}$ is greater than one.

Two conclusions thus emerge for the complete solution. One is that unless ab is less than one, any increase in the government expenditure term G would generate periodic fluctuations if the roots are complex. This was the reason why Phillips model introduced feedback control laws for choosing G such that the fluctuations in income could be reduced. Second, the stability condition can also be written as $b(1 - 1/k) < 1$, where k is the income multiplier and b the accelerator. The higher the value of k or, b the greater is the possibility of violating the stability condition. This provided the Keynesian policymakers an additional reason for influencing these parameter values through an effective government expenditure policy. The recent monetary theorists do not agree with such Keynesian prescriptions on several grounds e.g. (a) the parameters a, b are not that exactly known or estimable and (b) it ignores the expectation mechanism particularly of the monetary sphere. One type of expectation mechanism was formalized by the so-called Phillips curve or relation linking inflation and unemployment.

A linear continuous form of the Phillips curve is formalized as follows

$p = a - bU + c\pi$: sources of inflation
$D\pi = k_1(p - \pi)$: adaptive expectations   (9.3)
$DU = -k_2(m - p)$: effect of monetary policy.

Here $D = d/dt$ is the differential operator, p and π are the percentage increases in actual and expected price levels, U is the rate of overall unemployment, m is the percentage rate of growth of nominal money supply and all other variables are fixed positive parameters. The first relation of (9.3) implies that the rate of inflation measured by p is negatively related to unemployment (U) and positively related to the expected rate of inflation denoted by π. The second relation says that π rises (falls) whenever p is above (below) π i.e. it is much like a Walrasian adjustment for the expected rate of inflation. The third relation says that if the rate of growth of real money balance (= m-p) is positive, then the rate of unemployment declines through higher consumption demand.

Assuming m to be exogenously determined, we have three equations in three variables p, U and π, which by superposition theorem may be reduced to a single differential equation in a single variable. First, we eliminate p between the first relation of (9.3) to get

$$D\pi = k_1 a - k_1 b\, U + k_1(c - 1)\pi \qquad (9.4a)$$

Differentiating it,

$$D^2\pi = -k_1 b\, DU + k_1(c-1)\, D\pi$$
$$= (-k_1 b)\,[-k_2(m-p)] + k_1(c-1)\, D\pi \qquad (9.4b)$$

but $p = \pi + D\pi/k_1$ from the second relation of (9.3). On using this in (9.4b) we get the final reduced form of a second-order linear differential equation

$$D^2\pi + \{k_2 b + k_1(1-c)\}\, D\pi + (bk_1 k_2)\, \pi = bk_1 k_2 m.$$

The particular solution is $\pi^* = \pi_p = m$ i.e. if the particular solution is interpreted as the steady-state intertemporal equilibrium, then this equilibrium value of the expected rate of inflation depends on the rate of growth of nominal money supply which is largely controlled by the government monetary policy (e.g. Federal Reserve policy in U.S.).

The homogeneous part of the solution depends on the two roots of the characteristic equation $\lambda^2 + h_1 \lambda + h_2 = 0$, where $h_1 = k_2 b + k_1(1-c)$ and $h_2 = bk_1 k_2$

$$\lambda_1, \lambda_2 = \tfrac{1}{2}(-h_1 \pm \sqrt{h_1^2 - 4h_2}).$$

Again the various cases could be discussed. In particular, if $h_1^2 < 4h_2$ and $h_1 < 0$, then we would have damped periodic fluctuations in the expected rate of inflation. The time-paths of U and p can be easily derived from (9.4a) and the first relation of (9.3). Two points may be learned from this derivation. One is that a change in the autonomous policy regarding m (i.e. money supply) would affect the rate of price inflation and its expectation pattern. Second, the intensity of periodic fluctuations in π or, U can be influenced in part by a judicious monetary policy; however, this depends on the assumption of 'fine-tuning' i.e. the parameters are precisely known and the model is very realistic in modeling the aggregative behavior pattern.

We have so far discussed equilibrium Keynesian models. Consider now a disequilibrium Keynesian model where Y_d, Y_s denote the demand and supply sides of national output, I = real investment and G = net government expenditure

$$Y_d = (1-a)Y_s + I + G$$
$$I = v \cdot (DY_s) - h \cdot (DI) \qquad (9.5)$$
$$DY_s = k(Y_d - Y_s), \text{ if } Y_d \neq Y_s.$$

Here a, v, h, k are fixed positive parameters. Two features of this model are to be noted. First, the last equation specifies a process of market adjustment i.e. if demand exceeds supply and capacity is available, then output tends to rise; but if no capacity is available, then the increase is only in money value of output. The final reduced form equation is

$$D^2 Y_s + b \, DY_s - \frac{k}{h} G - kDG = 0$$

where D = d/dt and b = ak + 1/h - vk/h and DG = dG/dt assumes that government expenditure can vary over time. For this type of disequilibrium Keynesian model, Phillips discussed two situations, one where G is a constant so that DG is zero and the resulting path of $Y_s(t)$ has periodic fluctuations and a second situation where G is deliberately used as an anti-cyclical device - i.e. as a control variable. Three types of feedback policies: proportional ($G = f_1 Y_s$), derivative ($G = f_2 DY_s$) and integral ($G = f_3 \int Y_s \, dt$) were considered, where the feedback coefficients f_1, f_2 and f_3 are to be appropriately chosen so as to counteract the cyclical periodic fluctuations. Phillips derived the interesting result that sometimes such policies may themselves be destabilizing rather than stabilizing.

3.4 Nonlinear Dynamic Equations

Nonlinear differential and difference equations sometimes arise in economic problems due to the fact that many economic processes such as production have long gestation lags and kinks, which cannot be smoothed out very easily. Take the example of a coffee crop. It takes about 6 to 8 years for a coffee plant to mature and then it lasts for more than a decade. To describe their output behavior we may need a nonlinear growth curve. Take another example: the spread of a new innovation. When hybrid corn was first developed in U.S., it had a very slow rate of adoption by farmers, then it picked up very fast. The time profile of adoption of this new technology can only be adequately described by a nonlinear dynamic equation.

We would discuss two simple methods of nonlinear analysis. One uses a method of transformation by which the nonlinear equation is reduced to a linear one. This is however not always possible, since suitable transformations may not always exist.

The second method uses a linearizing technique for the nonlinear terms and tests for stability around certain points or paths. We would illustrate these methods by economic models and examples.

3.4.1 Logistic Model of Growth

The time profile of growth of many variables such as new technology, overall population and industrial production of such industries as computers, electronics, etc. follows very closely a logistic curve, which shows a very rapid initial growth rate, followed by a decline and an eventual asymptotic approach to some limit. The diffusion of newly available productive processes throughout an industry has been modeled by Edwin Mansfield in terms of a logistic growth model.

The nonlinear differential equation for a logistic model is of the form

$$Dy = ay(k - by) \tag{10.1}$$

where $y = y(t)$ is say population at time t, $D = d/dt$ and the other parameters are positive constants. It is clear that $Dy = 0$ in two cases: if $k = by$ and if $y = 0$. In the first case we get the upper asymptotic limit $y^* = y_U^* = k/b$. If we use the reciprocal transformation $x = 1/y$ for y not equal to zero, then $Dx = -Dy/y^2$ and (10.1) reduces to a first order linear differential equation with a constant non-homogeneous term

$$Dx = -ak\, x + ab.$$

The solution is

$$x = x(t) = (x_0 - \tfrac{b}{k}) \exp(-akt) + b/k$$

$$y = y(t) = 1/x(t) = \frac{k/b}{1 + B\exp(-akt)} \tag{10.2}$$

where x_0 is the initial value of $x = 1/y$ and $\exp(-akt) = e^{-akt}$ and $B = kx_0/b - 1$. It is clear from (10.2) that y reaches the upper asymptotic limit $y_U^* = k/b$ as $t \to \infty$.

In ecological models of a renewable resource like ocean fish, the logistic model appears as follows

$$Dy = f(y) - h(t)$$

where $\quad f(y) = ay(1 - y/k) =$ natural growth
$\quad h(t) = h_0\, u(t)\, y(t) =$ extraction rate.

This model assumes that the fish stock $y = y(t)$ grows as the logistic curve except for the extraction rate $h(t)$ which is proportional to $y(t)$ i.e. $h_0 =$ initial extraction rate, u(t) = rate of fishing effort. An optimal decision problem is how to determine the best rate u(t) of fishing effort over time?

3.4.2 Neoclassical Growth Model

This macro-dynamic model of economic growth due to Solow generalizes the Harrod-Domar version of the Keynesian growth model by providing a flexible production function with constant returns to scale, which permits capital-labor substitution through competitive factor markets. The aggregate production function is $Y = F(K,L)$, where Y, K, L are the aggregate levels of output, capital and labor in the whole economy. This function is assumed to be homogeneous of degree one i.e. subject to constant returns to scale. This implies that if each of the two factors, K and L are multiplied by a nonzero number m, then output Y is also multiplied by m i.e. $mY = F(mK,mL)$. Choose $m = 1/L$. Then $Y/L = F(K/L,1) = F(K/L)$. Denoting the per capita variables $y = Y/L$, $k = K/L$ we can write the neoclassical production function as

$$y = f(k), \quad MP_k = \partial y/\partial k > 0 \qquad \frac{\partial MP_k}{\partial k} = \frac{\partial^2 y}{\partial k^2} \leq 0$$

where the marginal product MP_k is assumed to be positive but nonincreasing ($\partial^2 y/\partial k^2 \leq 0$) as the capital-labor ratio k rises. We now introduce the equilibrium assumption that total savings ($S = sY$) equals total investment (I) where total investment equals additional capital installed (DK) plus the needs of depreciation ($R = rK$):

$$S = sY = I = DK + rK \qquad r > 0 \qquad (10.3)$$

where $D = d/dt$ and depreciation R is assumed to be proportional to K with a positive proportion r. On using the equilibrium condition (10.3) in the neoclassical production function $y = f(k)$ and noting the derivative of $k = K/L$

$$D\left(\frac{K}{L}\right) = Dk = \frac{DK}{L} - k\frac{DL}{L} = \frac{DK}{L} - kn$$

where labor L is assumed to grow at an exogenous rate $n > 0$, we obtain the capital-growth equation and the income growth equation as follows

$$Dk = sf(k) - (n + r)k$$
$$Dy = f_k\, Dk, \quad \text{where } f_k = \partial f(k)/\partial k. \qquad (10.4)$$

Equation (10.4) is sometimes called the fundamental growth equation of the neoclassical theory, or Solow equation. This has several interesting economic interpretations. First, the steady-state or particular equilibrium k* satisfies $Dk^* = 0$

thus implying $sf(k^*) = (n+r)k^*$. This says that savings per capita is just enough to finance the increase in total capital per head required by population growth (nk^*) plus depreciation (rk^*). Secondly, the growth-path of $k(t)$ in (10.4) is stable in the sense that if $k(t)$ exceeds (falls short of) the particular equilibrium value k^*, then $k(t)$ falls (rises) since $(n+r)k$ exceeds (falls short of) $sf(k)$. Just how long this process takes to converge to k^* is an important practical question and some computer simulations reported by Atkinson and others indicate that with sufficiently 'bad' parameter values, it may require decades. Thirdly, since per capita consumption (c) is $c = (1-s) f(k)$, the steady-state equilibrium condition (10.4) at $Dk = 0$ also implies

$$c^* = f(k^*) - (n+r)k^*.$$

Since $f(k^*)$ is a concave function of k^*, the equilibrium consumption level c^* can be maximized e.g. if $f(k^*)$ is twice differentiable, then the first order condition yields $f_{k^*} = n+r$ i.e. marginal product of capital must equal $(n+r)$. This type of results underlies much of the literature on optimal neoclassical growth.

We are of course interested in the nonlinearity of the differential equation (10.4). For specific results we assume a Cobb-Douglas type production function $Y = aK^b L^{1-b}$, which yields $y = f(k) = ak^b$, a and b being positive fractions as parameters. Hence we have to solve

$$Dk = sak^b - (n+r)k. \qquad (10.5)$$

This type of equation is known as the Bernoulli differential equation, which can be solved by a change of the dependent variable $x = k^{1-b}$, where b is the exponent of k in the production function. Then

$$Dx = (1-b)k^{-b} Dk. \qquad (10.6)$$

Multiplying both sides of (10.5) by $(1-b)k^{-b}$ and using the definition of x and (10.6) one gets

$$Dx = sa(1-b) - (n+r)(1-b)x$$

which is a linear first-order differential equation.

Its solution is

$$x(t) = [x_o - \frac{sa}{n+r}] \exp[-(n+r)(1-b)t] + \frac{sa}{n+r} .$$

Hence

$$k(t) = [x(t)]^{1/(1-b)}$$
$$= \{[x_0 - \frac{sa}{n+r}] \exp[-t(n+r)(1-b)] + \frac{sa}{n+r}\}^{1/(1-b)} \ .$$

It is clear that both $k(t)$ and $x(t)$ converge to k^* and x^* respectively, since $0 < b < 1$, where

$$x^* = sa/(n+r), \qquad k^* = (\frac{sa}{n+r})^{1/(1-b)} \ .$$

3.4.3 Lyapunov Theorem on Stability: Walrasian System

Lyapunov proposed two methods for solving nonlinear differential equations: (a) the indirect method where we try to explicitly solve the equation and test if the solution converges to an equilibrium value for $t \to \infty$, and (b) the direct method where we attempt to anlayze the stability of the differential equation system <u>without</u> <u>solving</u> it. In the first case, we can only obtain approximate solutions by applying suitable linearizations, if the system cannot be reduced to a linear form by change of variables. Hence we can only answer questions of <u>local stability</u> by this approach, i.e. when the initial values are sufficiently near the equilibrium point. For global stability it does not matter where the initial point is. Lyapunov's direct method is particularly useful for testing questions of global stability.

First, we describe the Lyapunov Theorem on global stability and then illustrate by a Walrasian model of general equilibrium. Consider a simultaneous system of autonomous differential equation in n variables $y_i(t)$, $i=1,2,\ldots,n$ with $D = d/dt$

$$Dy_i(t) = f_i(y_1, y_2, \ldots, y_n). \tag{11.1}$$

This system is autonomous because the right hand sides $f_i(\cdot)$ do not involve time t explicitly. Denote by the n-tuple vector y^* the equilibrium or particular solution at which $f_i(y^*) = 0$ for all $i=1,2,\ldots,n$. Then consider a continuous distance function $V = V(y - y^*)$ which measures the distance of the vector point y from y^*. For example the Euclidean distance function

$$V = +[\sum_{i=1}^{n} (y_i(t) - y_i^*)^2]^{\frac{1}{2}} \tag{11.2}$$

will serve the purpose, although other forms of distance may also qualify. If the distance function V defined e.g. in (11.2) satisfies the following three conditions:

  (i) V is positive if at least one of the quantities $y_1(t) - y_1^*$, $y_2(t) - y_2^*, \ldots, y_n(t) - y_n^*$ is different from zero, and V is zero if and only if each $y_i(t) =$

y_i^* for all $i=1,2,\ldots,n$

(ii) V goes to infinity as the norm $\|y(t) - y^*\| = [\sum_{i=1}^{n} (y_i(t) - y_i^*)^2]^{\frac{1}{2}}$ goes to infinity

(iii) $\frac{dV}{dt} = \sum_{i=1}^{n} \frac{\partial V}{\partial z_i} \frac{dz_i}{dt}$, $z_i = y_i(t) - y_i^*$ is negative if at least one of the n quantities $z_i = y_i(t) - y_i^*$, $i=1,2,\ldots,n$ is different from zero and dV/dt is zero, if and only if $y_i(t) - y_i^* = 0$ for all $i=1,2,\ldots,n$. Then the equilibrium point y^* with n components $y_1^*, y_2^*, \ldots, y_n^*$ is globally stable. Note that for the Euclidean distance function chosen in (11.2), the first two conditions are automatically satisfied. So only we have to test condition (iii) for checking global stability. As an example consider the stability of the van der Pol equation system used extensively in engineering dynamics

$$Dy_1 = y_2 + a\left(\frac{1}{3} y_1^3 - y_1\right)$$

$$Dy_2 = -y_1.$$

Choose the Lyapunov function $V = (y_1^2 + y_2^2)/2$ (note that we have not taken the positive square root of $(y_1^2 + y_2^2)$; this is all right because this V is continuous and satisfies the first two conditions (i), (ii) above). Then its derivative is

$$DV = dV/dt = y_1 Dy_1 + y_2 Dy_2$$

$$= ay_1^2 (y_2^2/3 - 1).$$

Hence if we choose a region where $y_1^2 < 3$, we have dV/dt negative. This suggests that we can define a region R in the two-dimensional space as $R = \{(y_1, y_2): y_1^2 + y_2^2 < 3\}$, and test if the three conditions above are satisfied. Condition (iii) is satisfied since we must have $y_1^2 < 3$ and also dV/dt is zero along the solution path only when y_1 is zero i.e. the point (y_1, y_2) must be on the y-axis. Conditions (i) and (ii) hold since in this region R we have $V \geq 0$ and $V \leq 3/2$. Hence the van der Pol equation is stable globally provided the equilibrium solution $y^* = (y_1^*, y_2^*)$ belongs to region R above.

Now consider the Walrasian general equilibrium system in n goods with prices p_i and excess demands $E_i(p) = [d_i(p_1, p_2, \ldots, p_n) - s_i(p_1, p_2, \ldots, p_n)]$. Let p^* be an equilibrium price vector with positive components p_i at which $E_i(p^*) = 0$ for all $i=1,2,\ldots,n$. The Walrasian tâtonnement process assumes a price adjustment process in terms of the following system of nonlinear differential equations

$$dp_i/dt = k_i E_i(p_1, p_2, \ldots, p_n), \qquad i=1,2,\ldots,n \tag{11.3}$$

where k_i's are positive constants representing speeds of adjustment in the different markets. Does there exist a region where the solution $p(t)$ of the nonlinear system (11.3) converges to the equilibrium price vector p^*? In general, the answer is obviously negative, unless suitable restrictions are placed on $E_i(p)$'s, since by making $E_i(p)$ discrete in time, the convergence of $p(t)$ to p^* may be prevented. Hence we make the following four assumptions on the excess demand functions:

(i) each $E_i(p)$ is single-valued and continuously differentiable,

(ii) it satisfies Walras' law i.e. $\sum_{i=1}^{n} p_i E_i(p) = 0$,

(iii) it is positively homogeneous i.e. $E_i(rp) = E_i(p)$ for all $i=1,2,\ldots,n$ and all positive scalars r, and

(iv) it satisfies the condition of gross substitutability i.e. $\partial E_i(p)/\partial p_j > 0$ for all p, where $i,j = 1,2,\ldots,n$ and $i \neq j$.

These conditions imply the important result

$$\sum_{i=1}^{n} p_i^* E_i(p) > 0 \qquad (11.4)$$

where p^* is an equilibrium price vector but p is not. This result (11.4) states that the so-called weak axiom of revealed preference holds for any pair of price vectors, one of which is an equilibrium price vector i.e. the weighted sum of excess demands evaluated at the equilibrium prices is strictly positive.

Now we define the Lyapunov distance function $V = \sum_{i=1}^{n} (1/k_i)(p_i - p_i^*)^2$, compute its derivative and use the adjustment process (11.3)

$$\frac{dV}{dt} = 2 \sum_{i=1}^{n} p_i E_i(p) - 2 \sum_{i=1}^{n} p_i^* E_i(p)$$

where the first term on the right-hand side is zero by Walras law, while the second term $2 \sum_{i=1}^{n} p_i^* E_i(p)$ is positive by (11.4) i.e. the weak axiom of revealed preference.

Hence dV/dt is negative for any p not equal to p^*, becoming zero only at the equilibrium point $p = p^*$. If p is positively proportional to the equilibrium price vector p^* i.e. $p = rp^*$, $r > 0$, then since the excess demands satisfy $E_i(rp) = E_i(p)$ we obtain $dV/dt = -2 \sum p_i^* E_i(p) = -2 \sum p_i^* E_i(rp^*) = - \sum p_i^* E_i(p^*) = 0$. Thus, by the Lyapunov Theorem the competitive Walrasian adjustment process is globally stable.

This result can be proved under more generalized conditions e.g. other types of distance functions may be used, differentiability of the excess demand functions may be relaxed.

3.4.4 Mixed Difference-Differential Equations

A very important class of nonlinearity arises in linear feedback policies, when there is discontinuous adjustment through lags in government stabilization policies. In production control systems described by linear differential equations, there may be a delay in applying the linear feedback control thus leading to a mixed difference-differential equation. Consider a macro-dynamic Keynesian model in equilibrium form

$$(aD + b)y(t) - g(t-s) = h(t)$$
$$g(t) = -ky(t) \quad (12.1)$$

under a Phillips-type proportional stabilization policy. Here $y(t)$ is real national income measured as deviations from some equilibrium level, $g(t)$ is government expenditure, $D = d/dt$, $h(t)$ is an autonomous component and the other parameters are positive constants. Note that if $s > 0$, there is a delay in government policy i.e. policy this month affects income s months later. This may also be due to the expectations formed by agents being affected by government policy. Note that if the time lag s is zero, then the characteristic equation has one negative real root, $-(b+k)/a$ only and hence the homogeneous solution is stable e.g. if $h(t)$ is constant h_o say, then the complete solution is

$$y(t) = (y_o - \frac{h_o}{b+k}) \exp(\lambda t) + \frac{h_o}{b+k} \ .$$

But in case of positive time lag ($s > 0$), the quasi-characteristic equation is

$$a\lambda + b + k \exp(-\lambda s) = 0 \quad (12.2)$$

This can now have complex roots, although for $s = 0$ only real roots are possible. The complex roots $\lambda = ix$, where $i = +\sqrt{-1}$ and x is a real number can be found from (12.2) by substitution

$$aix + b + k \exp(-isx) = aix + b + k (\cos sx - i \sin sx) = 0.$$

Separating the real and imaginary parts

$$b + k \cos sx = 0, \quad ax - k \sin sx = 0$$

we can analyze the stability or instability in the parameter space by choosing different nonnegative values of s and other parameters. This method of demarcating the parameter space as zones of stability and instability is called the method of D-partition. For example, if $b > 0$ and $k = 0$ then for any s, the quasi-polynomial equation (12.2) has no roots with positive real parts. Hence this specifies a region of asymptotic stability. On passing from this region to the next region across the straight line b + k = 0, one root with a positive real part appears, thus indicating a zone of instability.

The same situation would arise in a second order multiplier-accelerator model

$$(aD^2 + bD + c) y(t) - g(t-s) = h_o$$
$$g(t) = -ky(t)$$

when there is a lag in government policy. Thus the Phillips-type model emphasized the point that the government policy itself may generate periodic fluctuations and this is more likely when we have a multisectoral linear dynamic model where some parameters are either incompletely known or partly random.

The analysis above also suggests the potential dangers of drawing wrong conclusions by linearly approximating a nonlinear differential equation. To take another example, consider the van der Pol equation discussed before

$$D^2 y - a(1 - y^2)Dy + y = 0 \qquad (12.3)$$

This is nonlinear when a is nonzero. However, if the absolute value of y i.e. $|y|$ is small and y^2 may be neglected in comparison with unity, then the equation (12.3) can be approximately written as

$$D^2 y - aDy + y = 0 \qquad (12.4)$$

which is a linearized form of van der Pol's equation. Its solution is

$$y = y(t) = A_1 \exp(\lambda_1 t) + A_2 \exp(\lambda_2 t) \qquad (12.5)$$

where A_1, A_2 are the two constants of integration to be determined by initial conditions and λ_1, λ_2 are the two roots of the characteristic equation

$$\lambda^2 - a\lambda + 1 = 0. \qquad (12.6)$$

If a > 0, then the two roots of (12.6) evidently have positive real parts. The linearized system (12.4) is thus unstable. This may also be easily verified by Lyapunov Theorem. However, if a = 0, then the equation (12.6) has imaginary (i.e. complex) roots and therefore no conclusions as to the stability of the nonlinear system may be drawn from the behavior of the linearized system. Nevertheless, it is clear that a system governed by (12.3) will be stable in this case, since it has a bounded but periodic solution of the form y = A sin(t + θ) when a = 0. It should be noted that this periodic fluctuation differs from the limit cycle oscillation which occurs when a > 0 in that the amplitude term A can be made arbitrarily small by choosing the initial conditions appropriately.

Finally, if a < 0, the roots of (12.6) have negative real parts, indicating that the linearized system is stable. One can also show directly by using Lyapunov's Theorem that the nonlinear van der Pol equation (12.3) is stable when the parameter a is negative.

Thus we conclude that nonlinear dynamic systems need more careful investigation when stability questions arise. It is only recently that nonlinear dynamic models are being applied both theoretically and empirically in economic systems.

PROBLEMS THREE

1. Metzler's inventory cycle model for the whole economy assumes the following relations for national income $Y(t)$, expected and realized sales $\hat{S}(t)$, $S(t)$ and the desired and actual inventory levels $\hat{H}(t)$, $H(t)$

$$Y(t) = \hat{S}(t) + [\hat{H}(t) - H(t-1)] + I_o$$
$$\hat{S}(t) = S(t-1); \quad S(t) = b\, Y(t), \quad 0 < b < 1$$
$$\hat{H}(t) = k\, \hat{S}(t), \quad k > 0; \quad I_o = \text{autonomous investment}.$$

Show that if $b < (1+k)^{-1}$, then the income path has periodic fluctuations but convergent, while if $b > (1+k)^{-1}$ the path is divergent.

2. Show that the general solution of the homogeneous system of differential equations

$$\begin{pmatrix} D\, x_1 \\ D\, x_2 \end{pmatrix} = \begin{bmatrix} -1 & 1 \\ -5 & 3 \end{bmatrix} \begin{pmatrix} x_1 \\ x_2 \end{pmatrix}$$

is given by

$$\begin{pmatrix} x_1(t) \\ x_2(t) \end{pmatrix} = c_1 \begin{pmatrix} \cos t \\ 2\cos t - \sin t \end{pmatrix} e^t + c_2 \begin{pmatrix} \sin t \\ 2\sin t + \cos t \end{pmatrix} e^t$$

where c_1, c_2 are two constants to be determined.

3. The limit pricing model assumes that the rival firms with output x_2 will enter the market as follows

$$D x_2(t) = k\, (p(t) - p_o), \quad k > 0; \quad p_o > 0$$
$$p(t) = a - b\, (x_1(t) + x_2(t)); \quad a, b > 0$$

whenever the dominant firm with output $x_1(t)$ charges a price $p(t)$ much above the competitive level p_o (fixed). What would be the time path and its stability of output $x_2(t)$, if the dominant firm adopts the following three policies

(a) $x_1(t) = \bar{x}_1 > 0$ fixed
(b) $x_1(t) = h_1 x_2(t) + h_2\, D\, x_2(t)$
    $h_1, h_2 > 0$ and $D = d/dt$
(c) $x_1(t) = h\, x_2(t-1), \quad h > 0$.

4. Consider a linear optimization problem in a scalar variable x and its dual y. max z = cx subject to ax = 1, where the Lagrangian function is $L = L(x,y) = cx + y(1 - ax)$. Suppose we have not reached the optimal solution and the gradient method of computation is adopted i.e. x and y are revised as follows

$$dx/dt = k_1 L_x; \quad k_1 > 0, \quad L_x = \partial L/\partial x = c - ay$$
$$dy/dt = -k_2 L_y; \quad k_2 > 0, \quad L_y = \partial L/\partial y = 1 - ax.$$

Show that the iterative calculations follow periodic fluctuations in its computational path.

5. Let y(t) be the temperature of an oven at time t measured in suitable units and x(t) be the rate of fuel supply (i.e. the only control variable) to the oven in the interval [t-1,t] measured in such units that we can write

$$y(t) = a\, y(t-1) + x(t) + u(t)$$

where u(t) is a random variable with zero mean and variance unity for all t and a is a fixed constant such that $0 < a < 1$. Suppose we observe past values of output i.e. y(t-1), y(t-2),... to determine present action or control x(t) in such a way that the future output is as near as possible to the desired output level denoted by $\bar{y}$ (fixed). How are we to choose the values of x(t)?

6. Analyze the stability of the Solow growth model

$$Dk = sf(k) - (n+r)k$$
$$k(0) = k_0 > 0 \text{ given}$$

when f(k) may take any of the three forms
(a) $f(k) = ak(t), \quad a > 0$
(b) $f(k) = a + b_1 k(t) - b_2 k^2(t); \quad a, b_1, b_2 > 0$
(c) $f(k) = a_1 k(t)$ for $k(t) \leq 100$ but, $f(k) = a_2 k(t)$ for $k(t) > 100$ where a_1, a_2 are positive constants.

7. A two-sector investment planning model due to Mahalanobis postulates that the outputs of investment I(t) and consumption goods C(t) will grow according to the following equations

$$I(t) = I(t-1) + u_i\, b_i\, I(t-1)$$
$$C(t) = C(t-1) + u_c\, b_c\, I(t-1)$$
$$I(0) = I_0, \quad C(0) = C_0 \text{ are given}$$

where u_i, u_c are the nonnegative proportions of investment fund $I(t-1)$ available to the planner such that $u_i + u_c = 1$ and b_i, b_c are positive constants representing the productivity of investment in the two sectors. Compute the growth path of national income $Y(t) = C(t) + I(t)$ for the following values of u_i, u_c

(i) $u_i = u_c = 1/2$; (ii) $u_i = 0.9$, $u_c = 0.1$.

Compute the percentage growth rates of $I(t)$ and $I(t)$. Do they converge?

8. Consider the macro-dynamic Keynesian model subject to economic regulation by government policy

$$dx_2/dt = -b_1 x_1 + m x_3 + m g$$
$$dx_1/dt = x_2; \quad dx_3/dt = b_2 x_2 - b_3 g + b_4 x_1$$

where all variables x_i are in per capita terms: x_1 = real national income, x_3 = gross investment and the positive parameters are $b_1 = ms+n$, $b_2 = v/k$, $b_3 = n+1/k$, $b_4 = vn/k$ where s = saving coefficient, n = proportional rate of growth of labor force (given), v, k = parameters of the induced investment function (i.e. accelerators) and m = coefficient of output adjustment. Discuss the stability of the growth path of national income $x_1(t)$ under the following three Phillips-type regulatory policies:

(a) proportional: $g = f_1 x_1$
(b) derivative: $g = f_2 (dx_1/dt)$
(c) integral: $g = f_3 \int_0^t x_1 \, dt$

where f_1, f_2, f_3 are fixed constants.

9. A commonly used dynamic equation for the fishing population (e.g. tuna) in ocean $x(t)$ is of the type

$$x(t+1) = x(t) + r \, x(t) \, [1 - \frac{x(t)}{k}] - u(t)$$

where r is a fixed groth rate parameter (r > 0) and k is a positive parameter denoting unfinished equilibrium of $x(t)$ and $u(t)$ is the rate of catch. If the rate of catch follows a proportional rule: $u(t) = a + b \, x(t)$, a and b being positive constants, what would be the time path of $x(t)$ given an initial level $x(0)$?

10. Consider the nonlinear differential equation

$$D^2 y(t) + w^2 y(t) + b y^3(t) = 0$$

known as the Duffing equation. If b is very very small, then the term $by^3(t)$ depicts a small deviation from linearity. Test by Lyapunov Theorem if the solution is stable when b is zero. If b is positive, then the above equation is called a <u>hard spring</u> case, whereas the case b < 0 depicts the movement of a <u>soft</u> spring. Test by Lyapunov Theorem if there is a region where the solution is stable for the hard and soft spring cases.

CHAPTER FOUR
DYNAMIC OPTIMIZATION AND CONTROL

Dynamic optimization is most fundamental to our understanding of economic behavior analyzed through economic models. This is so for several reasons. For one thing, it includes static optimization as a special case, since a dynamic model in its steady-state equilibrium may represent a static equilibrium model. Second, many economic decisions e.g. investment decisions have long run consequences, so that a policy which is optimal in the sense of maximizing short run profits for a firm may not be optimal in the long run, since it may invite more entry and thereby reduce long run profits. This apparent inconsistency between the short-run or myopic policy and the long run policy may lead to various adjustments in firm behavior, particularly when the future is only incompletely known or imperfectly forecast. Third, the understanding of <u>dynamic</u> or <u>intertemporal</u> or <u>moving equilibrium</u> as an optimum concept, whenever such an interpretation is meaningful in economic models, provides us with a normative yardstick against which actual realized behavior can be compared and tested for correspondence. This is particularly important for differential game models, which are increasingly applied in recent literature, where two or more players seek to reach an intertemporal equilibrium or optimum by appropriately choosing their own strategies over time. Lastly, dynamically optimal policies are most important for applying econometric models over time, since updating of policy may be performed as more data become available over time. In the recent literature on economic applications of control theory, these policies are studied under models of optimal, adaptive and stochastic control.

For dynamic economic systems governed by difference and differential equations, there exist four general methods of dynamic optimization: nonlinear programming, calculus of variations, Pontryagin's maximum principle and Bellman's dynamic programming. The first is most suitable for the dynamic optimization subject to difference equation models; the second for differential equation models with no inequality constraints on the control or state variables. The third is suitable for both discrete time (i.e. discrete maximum principle) and continuous time (i.e. continuous maximum principle) dynamic equations when there may exist inequality constraints on the state or control variables. Lastly, the algorithm of dynamic programming is applicable most generally provided some conditions are fulfilled. Our objective is to illustrate these techniques by simple and applied economic models, emphasizing in particular Pontryagin's maximum (or minimization) principle which has found widemost applications in current economic literature.

4.1 Intertemporal Optimization: Discrete Time

One of the simplest intertemporal optimization problems occurs when we have a dynamic economic system governed by linear difference equations, and we have two sets of time-dependent variables called the state variables and the control (i.e. decision) variables and the objective of the policymaker is to choose the control variables so that an intertemporal objective function is optimized. Thus, if the firm is the policymaker choosing investment strategies as its control, its objective may be to maximize instead of the current short run profits a discounted sum of future profits over a planning horizon. We illustrate by some simple economic examples.

4.1.1 Competitive Model with Inventories

A single product firm under perfect competition is facing demand $d(t)$, when its expected price is $p(t)$. The output $u(t)$ and inventories $x(t)$ are related to expected demand $d(t)$ as

$$x(t) = x(t-1) + u(t) - d(t) \qquad t=1,2,\ldots,T$$
$$x(0) = x_0 \text{ given} \qquad x(t), u(t), d(t) \geq 0, \text{ all } t. \qquad (1.1)$$

We assume for simplicity that the firm has a planning horizon of T periods with no discounting, in which total production costs $c(u(t))$ and total inventory costs $c(x(t))$ are assumed unchanged over time, although expected prices $p(t)$ may vary over time. As a price taker the firm has no influence on the price $p(t)$. Total profits for the discrete horizon is then

$$z(T) = \sum_{t=1}^{T} [p(t) d(t) - c(u(t)) - c(x(t))] \qquad (1.2)$$

The firm's decision problem is to maximize total long run profits $z(T)$ over the planning horizon by choosing the optimal values of output and the associated values of inventories i.e. $u(t), x(t), t=1,2,\ldots,T$ over the entire future path, given the initial level x_0 of inventories and an estimate or forecast of demand $d(t)$ over the horizon $1 \leq t \leq T$.

Note two important features of this dynamic optimization problem. First, we are solving for optimal values of $\{u(1), u(2),\ldots,u(T)\}$ and the associated values of $\{x(1), x(2),\ldots,x(T)\}$ i.e. the time paths of output and inventories and not a single point output and inventories. Second, the dynamic elements enter first through the difference equation of (1.1) and then through the long run objective function which is merely the sum of short run profits over the future horizon. Here $x(t)$ is usually called the state (or output) variable, $u(t)$ the control (or decision) variable and $p(t)$, $d(t)$ the exogenous variables that have to be forecast. The difference equation (1.1) cannot be solved unless the output rate $u(t)$ is specified.

In feedback control the output rates are fixed in relation to the past levels of inventories and demand e.g. $u(t) = -ax(t-1)$, $a > 0$ in proportional controls, as we have seen before in Chapter Three in multiplier-accelerator and inventory models. In optimal control, we determine the unknowns $\{u(t), 1 \leq t \leq T\}$ by maximizing long run profits $z(T)$ defined in (1.2).

The nonlinear program (NLP) max $z(T)$ subject to (1.1) may be solved here directly by substituting $d(t) = x(t-1) - x(t) + u(t)$ in the objective function (1.2) and applying the Kuhn-Tucker necessary conditions:

$$\partial z(t)/\partial u(t) \leq 0, \quad u(t) \geq 0; \quad t=1,2,\ldots,T$$
$$\partial z(t)/\partial x(t) \leq 0, \quad x(t) \geq 0; \quad t=1,2,\ldots,T \quad (1.3)$$

where
$$z(t) = p(t)\, d(t) - c(u(t)) - c(x(t))$$
$$= p(t)\{x(t-1) - x(t) + u(t)\} - c(u(t)) - c(x(t)).$$

This yields

$$p(t) - c'(u(t)) \leq 0, \quad u(t) \geq 0, \quad \text{all } t$$
$$p(t+1) - p(t) - c'(x(t)) \leq 0, \quad x(t) \geq 0 \quad \text{all } t$$

where prime denotes partial derivatives e.g.

$$c'(u(t)) = \partial c(u(t))/\partial u(t), \quad c'(x(t)) = \partial c(x(t))/\partial x(t).$$

These necessary conditions (1.3) are also sufficient if the function $z(t)$ is concave in $u(t)$ and $x(t)$ i.e. the cost functions $c(u(t))$, $c(x(t))$ are convex and differentiable.

To obtain some specific results, assume for simplicity that the optimal time paths $\{u(t), x(t); 1 \leq t \leq T\}$ are such that for each t, $u(t)$, $x(t)$ are positive. Then we have from (1.3) by Kuhn-Tucker Theorem

$$p(t) = c'(u(t)), \quad t=1,2,\ldots,T$$
$$p(t+1) - p(t) = c'(x(t)), \quad t=1,2,\ldots,T-1 \quad (1.4)$$
hence $\quad c'(u(t)) + c'(x(t)) = c'(u(t+1)), \quad t=1,2,\ldots,T-1.$

The first optimal condition of (1.4) states that the expected price at time t must equal marginal cost at time t; the second condition is that the marginal storage cost must equal the change in expected prices between the two consecutive periods. The third condition at the optimal profit says that it is never worthwhile to produce in period $t+1$ an amount whose marginal production cost exceeds the sum of marginal production and storage costs in the previous period.

Assume further that the cost functions are quadratic and convex

$$c(u(t)) = a_1 u(t) + a_2 u^2(t) \qquad a_1, a_2 > 0$$
$$c(x(t)) = b_1 x(t) + b_2 x^2(t) \qquad b_1, b_2 > 0. \qquad (1.5)$$

Then we obtain from (1.4) and (1.1) after some algebraic manipulation the following second-order difference equation in p(t)

$$p(t+1) - k\, p(t) + p(t-1) = -(a_2\, d(t) + 2a_1 b_2^2)/2a_2\, b_2 \qquad (1.6)$$

which has the characteristic equation

$$\lambda^2 - k\lambda + 1 = 0, \qquad k = 2 + (b_2/a_2). \qquad (1.7)$$

Its two roots

$$\lambda_1, \lambda_2 = (1/2)\,(k \pm \sqrt{(k^2 - 4)})$$

are real and distinct since $k > 2$.

Note that the product of the two roots $(\frac{k + \sqrt{k^2-4}}{2})(\frac{k - \sqrt{k^2-4}}{2})$ equals one, hence each root is the reciprocal of the other. On using the two roots of (1.7) the complete nonhomogenous difference equation (1.6) can be solved for p(t) and this solution then may be used to specify the optimal time-paths of u(t) and x(t)

$$u(t) = (p(t) - a_1)/2a_2$$
$$x(t) = (p(t+1) - p(t) - b_1)/2b_2.$$

The initial condition $x(0) = x_0$ and a fixed terminal condition say $x(T) = x_T$ provide the two independent constants in the solution of the difference equation.

This problem has the following economic features that the expected price cannot be used by the firm as a control variable, since it is a price taker under a competitive model; also demand facing the individual firm is by assumption independent of the expected price. A contrasting example is that of a single firm monopoly. To illustrate assume there is no inventories, the production cost is linear: $c(u(t)) = a\, u(t)$ and the demand function is of the form

$$d(t) = h_1 - h_2\, p(t) - h_3\, [p(t) - p(t-1)]$$

where the positive parameter h_3 shows the adverse effect of price rise on demand.

Since there is no inventories the strategy of the monopolist is to produce $u(t)$ equal to demand $d(t)$ and set the price $p(t)$ in such a fashion that total profit is maximized. Current profit at time t is

$$z(t) = (p(t) - a) d(t)$$
$$= -ah_1 + p(t) [a(h_2 - h_3) + h_1] - (h_2 - h_3) p^2(t)$$
$$- h_3 p(t) p(t-1) \qquad (1.8)$$
$$z(T) = \sum_{t=1}^{T} z(t) = \text{long run profits.}$$

The price variable now is the only control variable. Applying the Kuhn-Tucker Theorem and assuming that the nonnegativity condition $p(t) \geq 0$ is fulfilled for all t, we obtain $\partial z(T)/\partial p(t) = 0$ thus yielding

$$h_3 (p(t+1) + p(t-1)) + 2(h_2 - h_3) p(t) - a(h_2 - h_3) = 0$$

The solution of this second order difference equation is given by

$$p(t) = A_1 \lambda_1^t + A_2 \lambda_2^t + p^* \qquad (1.9)$$

where $p^* = a(h_2 - h_3)/2h_2$ is the particular solution ($h_2 > h_3$ is needed for p^* to be positive), A_1, A_2 are the two constants determined by the two boundary conditions e.g. $p(0) = p_0$ and $p(T) = p_T$ say and λ_1, λ_2 are the two roots of the characteristic equation

$$\lambda^2 + [2(h_2 - h_3)/h_3] \lambda + 1 = 0.$$

Assuming real and distinct roots (i.e. the case $h_2 > 2h_3$) where $\lambda_1 = 1/\lambda_2$, the complete optimal trajectory of price can be written as (1.9) with A_1, A_2 solved from

$$p_0 = A_1 + A_2 + p^* \text{ and } p_T = A_1(\lambda_1)^T + A_2(\lambda_2)^T + p^*$$

i.e.

$$A_1 = [(\lambda_2)^T - (\lambda_1)^T]^{-1} [(p_0 - p^*)(\lambda_2)^T - (p_T - p^*)]$$
$$A_2 = [(\lambda_2)^T - (\lambda_1)^T]^{-1} [(p_T - p^*) - (\lambda_1)^T (p_0 - p^*)]$$

Once the optimal time path of $p(t)$, $1 \leq t \leq T$ is found, the associated optimal trajectory of output $u(t) = d(t)$ is easily found by substitution.

4.1.2 Stabilization in Commodity Markets

Most studies of price stabilization policies in world commodity markets either attempt to solve for the socially optimal level of storage by assuming a buffer fund agency or compute the competitive level of storage. However, for many primary commodities e.g. coffee, cocoa the world market is highly imperfect, as the share of large producers is very dominant. It is possible to illustrate the optimal storage rules under competitive versus cartelized commodity markets in terms of the model we outlined in the earlier section.

Assuming all firms are alike, the optimizing decision of a representative competitive firm is obtained by maximizing a discounted sum of future profits:

$$\text{Max } z(T) = \sum_{t=1}^{T} r^t [p(t)d(t) - c(u(t)) - c(x(t))]$$

subject to $x(t) = x(t-1) + u(t) - d(t),$ \hfill (2.1)

$x(0) = x_0$ (given) \hfill $t=1,2,\ldots,T$

where $r = 1/(1 + i)$ is the discount factor to apply to next year's sale when the interest rate is i. For simplicty we assume $c(x(t)) = b\, x(t)$, b is the constant marginal storage cost. Since future harvest or production is uncertain, the variable u(t) has to be interpreted as expected output. If the competitive producers are risk neutral, they do not pay any attention to the variance or random fluctuations of harvest and its effect on profit. Hence, one can solve the model (2.1) above by following the Kuhn-Tucker Theorem as before. This yields the optimal decision rule to be followed by the competitive producers in the aggregate

$p(t) + b \geq r\, E\{p(t+1)\}$ \qquad $x(t) \geq 0$ \hfill (2.2)

$p(t) - c'(u(t)) \leq 0$ \qquad\qquad $u(t) \geq 0.$ \hfill (2.3)

Here E denotes expectation and this is explicitly introduced here to indicate the prominent role of the need to forecast future prices. If the producers are not risk neutral but risk averse, a separate cost due to forecasting errors will be added to this term. Note that the optimality condition (2.2) says that the risk neutral competitive equilibrium level of storage will be to store enough to drive up today's price and drive down next year's expected price until there exist no further arbitrage gains to be made on further storage.

Now assume that the commodity market is imperfect, so that the producers are not price takers. Define total expected revenue by $R(t) = p(t)d(t)$ and the marginal revenue by $m(t) = \partial R(t)/\partial u(t)$. The optimal decision rules now become

$m(t) + b \geq rE\{m(t+1)\}$ \qquad $x(t) \geq 0$ \hfill (2.4)

$m(t) - c'(u(t)) \leq 0$ \qquad\qquad $u(t) \geq 0.$ \hfill (2.5)

Here we assume that the expected demand function is linear in price and the dominant producers do all the storage. Whether or not dominant producers will tend to stabilize prices more than the competitive level depends on the way the elasticity of demand varies with price, since this will determine the shape of the marginal revenue schedule relative to the demand schedule. It is clear that if demand is linear, then dominant producers will undertake significantly more storage than the competitive producers for two reasons. Price stability may increase the market share of the dominant producers and they may be more risk averse in the sense of higher sensitivity to high variance of profits.

For instance, if an additional cost $c(v(t))$ due to the variance $v(t)$ of total revenue $R(t)$ is subtracted in the discounted profit function, the optimal decision rules would be

$$m(t) + b + \frac{\partial c(v(t))}{\partial x(t)} \geq rE\{m(t+1)\} \qquad x(t) \geq 0$$

$$m(t) - c'(u(t)) - \frac{\partial c(v(t))}{\partial u(t)} \leq 0 \qquad u(t) \geq 0.$$

This leaves us with two conclusions. One is that the presence of imperfect competition in world commodity markets tends to reduce price instability measured by the price variance and the more risk averse the dominant producers are, the more important would be the price stabilizing impact.

4.1.3 Optimal Growth in Neoclassical Models

The Solow equation of neoclassical growth model we considered in Chapter Three is of the form

$$Dk(t) = \frac{dk}{dt} = s\, f(k) - (n+r)k. \tag{3.1}$$

By defining per capita real consumption as $c = (1-s)\, f(k) = f(k) - sf(k) =$ output minus saving, we obtain the discrete time version of the time path of $c = c(t)$

$$c = f(k) - (n+r)\, k(t) - (k(t+1) - k(t)) \tag{3.2}$$

where Dk is approximated by $k(t+1) - k(t)$. Assume a social welfare functional in the form of a discounted sum of utilities $U(c)$ of per capita real consumption

$$J = \sum_{t=0}^{T} \delta^{-t}\, U(c) \tag{3.3}$$

where $\delta = 1+i$ is the discount rate with $i =$ positive rate of interest and $U(c)$ is the utility function assumed to be concave and differentiable. Optimal growth models usually assume that the aim of the society (or the social planner) is to maximize J in (3.3) subject to (3.2) when the initial k_o and terminal values k_{T+1} of capital stock are prescribed. This type of problem has been called the generalized Ramsay problem after Frank Ramsay. As before we apply the Kuhn-Tucker necessary condition to obtain

$$U'(c(t)) [1 + f'(k(t))] - (1 + n + r)\delta U'(c(t-1)) = 0 \qquad (3.4)$$

where prime denotes the respective partial derivative. This necessary condition is also sufficient by the assumption of concavity of $U(c(t))$. A particular solution k^* of the difference equation and the associated value $c^* = c(t) = c(t-1)$ for all t yields

$$1 + f'(k^*) = (1 + n + r)\delta \qquad (3.5)$$

since $U'(c(t)) = U'(c^*) = U'(c(t-1))$. Samuelson has distinguished between three types of golden rules

(a) Ramsay rule when $r = n = 0$, $\delta = 1$ and the long run equilibrium level k^*, also called the bliss level satisfies

$$f'(k^*) = 0$$

(b) Phelps-Swan rule when $r = 0$, $\delta = 1$ with $f'(k^*) = n$ i.e., marginal productivity of capital equals the growth rate of labor, and

(c) Schumpeter-Ramsay bliss level when $1/\delta = (1+n)$ and $r = 0$:

$$f'(k^*) = 0.$$

Any solution satisfying the optimality condition (3.4) is often called the turnpike and the following convergence theorem is often called the turnpike theorem in one of its simpler versions: Given the initial and terminal capital stocks k_o, k_{T+1}, the optimal trajectory (or time path) of capital $k(t)$ obtained by maximizing J in (3.3) will arch toward the turnpike defined by the equation (3.4) and spend most of its time in a near neighborhood of that turnpike before finally moving to the prescribed terminal capital stock.

However interesting the particular solution k^*, c^* may be, its stability depends on the complete solution of the system (3.4) i.e., the particular solution is only a steady-state equilibrium which may sometimes be reached only in infinite time. To show the path of convergence we assume specific functional forms for $U(c)$ and $f(k)$ as follows

$$U(c) = U(c(t)) = c(t) - (a/2) c^2(t) \qquad a > 0$$
$$f(k) = bk. \qquad b > 0 \qquad (3.6)$$

Since we require marginal utility $\partial U/\partial c = 1 - ac(t)$ to be positive, we impose the condition $c(t) < 1/a$ for all t by choosing our scale and units of measurement.

On substituting (3.6) into (3.4) and using the relation $c(t) = bk(t) + (1-n-r)k(t) - k(t+1) = gk(t) - k(t+1)$, $g = 1+b-n-r$, we obtain a second order difference equation with constant coefficients

$$h_1 + (h_3 + gh_2) k(t) + h_2(t+1) + gh_3 k(t-1) = 0$$

where
$$h_1 = 1 + b - \delta(1+n+r), \qquad h_2 = a(1+b)$$
$$h_3 = a(1+n+r)\delta.$$

The characteristic equation is

$$h_2 \lambda^2 + (h_3 + gh_2)\lambda + gh_3 = 0$$

which has two real and distinct roots if gh_2 is not equal to h_3. These roots are $\lambda_1 = g$, $\lambda_2 = h_3/h_2$ and each root lies between zero and one, if $0 < g < 1$, $0 < h_3/h_2 < 1$. Hence the complete solution is

$$k(t) = A_1 \lambda_1^t + A_2 \lambda_2^t + k^*$$

where A_1, A_2 are determined by the initial (k_0) and terminal (k_{T+1}) values i.e.

$$\begin{pmatrix} A_1 \\ A_2 \end{pmatrix} = \begin{bmatrix} 1 & 1 \\ \lambda_1^{T+1} & \lambda_2^{T-1} \end{bmatrix}^{-1} \begin{pmatrix} k_0 - k^* \\ k_{T+1} - k^* \end{pmatrix}.$$

If each root λ_i is a proper fraction i.e. $0 < |\lambda_i| < 1$, i=1,2 then it is clear that as $t \to \infty$, $k(t)$ will converge to k^*. Then the optimal consumption path $c(t)$ will converge to the steady-state equilibrium $c^* = (b-n-r)k^*$ where the condition $b > n+r$ is required for positivity of c^*, assuming k^* to be positive.

4.1.4 Discrete Pontryagin Principle

Consider the competitive model (1.1), (1.2) in a slightly different form, which is otherwise known as the optimal production-scheduling model due to Holt, Modigliani, Muth and Simon, also called the HMMS model. Here the firm uses a forecast $\hat{d}(t)$ of demand $d(t)$ to minimize the total cost $C(T)$ over the planning horizon $t = 0$ to $t = T-1$ i.e.

$$\text{Min } C(T) = \sum_{t=0}^{T-1} [c(u(t)) + c(x(t))] \tag{4.1}$$

subject to
$$x(t+1) = x(t) + u(t) - \hat{d}(t)$$
$$x(0) = x_0 \text{ given.} \tag{4.2}$$

Assume quadratic and convex production and storage costs e.g. $c(u) = a_1 u(t) + a_2 u^2(t)$, $c(x) = b_1 x(t) + b_2 x^2(t)$. Define the Lagrangean function

$$L = \sum_{t=0}^{T-1} [c(u) + c(x) + y(t+1)(x(t) + u(t) - \hat{d}(t) - x(t+1))] \tag{4.3}$$

where $y(t+1)$ is the Lagrange multiplier associated with the difference equation (4.2). Assuming interior solutions i.e., $u(t) > 0$, $x(t) > 0$ for all t, the first order conditions for the minimum are applicable i.e. $\partial L/\partial u(t) = 0$, $\partial L/\partial x(t) = 0$. These yield for $t = 0,1,2,\ldots,T-1$

$$\begin{aligned} a_1 + 2a_2 u(t) + y(t+1) &= 0 \\ b_1 + 2b_2 x(t) + y(t+1) - y(t) &= 0 \end{aligned} \tag{4.4}$$

also $\quad x(t) + u(t) - \hat{d}(t) - x(t+1) = 0.$

By eliminating $x(t)$ and $u(t)$ in the system (4.4) we obtain for the optimal trajectory the time path of $y(t)$ which is nothing but the dynamic shadow price

$$y(t+2) - \frac{b_2 + 2a_2}{a_2} y(t+1) + y(t) - (\frac{a_1 b_2}{a_2} + 2b_2 \hat{d}(t)) = 0. \tag{4.5}$$

The characteristic equation for the homogeneous part of this second-order difference equation is

$$\lambda^2 - k\lambda + 1 = 0, \quad k = (b_2 + 2a_2)/a_2.$$

Clearly, the product of the two roots is unity, so one root is the reciprocal of the other.

We have to take care of the two boundary conditions now. One is given by the initial value x_0 but since the Lagrangean function L in (4.3) involves $x(T)$ when $t = T-1$ we have to set $\partial L/\partial x(T)$ equal to zero, when $x(T)$ is free but T is fixed. Thus

$$\partial L/\partial x(T) = y(T) = 0$$

is called the transversality condition. We have to solve the system of difference equations (4.4), the reduced form of which is (4.5) subject to fixed values of $x(0)$ and $y(T) = 0$. This is therefore called a two-point, boundary-value problem.

Instead of the Lagrangean function, Pontryagin's minimum/maximum principle (here it is the minimum because the objective function is one of minimization) defines a function called the Hamiltonian H and states the necessary conditions in two parts. For our example $H = \sum_{t=0}^{T-1} H(t)$ where

$$H(t) = c(u(t)) + c(x(t)) + y(t+1)(x(t) + u(t) - \hat{d}(t))$$

The first part of the necessary condition, called the adjoint equation for $t = 0,1,\ldots,T-1$ are

$$\begin{aligned} x(t+1) - x(t) &= \partial H(t)/\partial y(t+1) \\ y(t+1) - y(t) &= -\partial H(t)/\partial x(t). \end{aligned} \quad (4.6)$$

The second part says that the current value Hamiltonian function H(t) must be a minimum for every discrete time t with respect to the control variable u(t) i.e. in this case

$$\partial H(t)/\partial u(t) = 0. \quad (4.7)$$

It is clear that the two necessary conditions (4.6), (4.7) are nothing but (4.4). Further we have the optimality condition for the terminal time T if x(T) is free to vary

$$h(T)\, y(T) = 0 \quad (4.8)$$

where $h(t)$ is a small linear perturbation of the optimal trajectory i.e. $h(t) = [\tilde{x}(t) - x(t)]/\alpha$ where α is a small positive constant and $\tilde{x}(t)$ is a path close to the optimal path $x(t)$.

Note that if there are additional inequality constraints on the control variable u(t) or the state variable x(t), these have to be incorporated in solving the adjoint equations and (4.7) and (4.8).

Pontryagin's minimum principle has two advantages over the Kuhn-Tucker theory applied to optimal control problems. One is that it separates the necessary conditions into two parts. This makes it a very convenient computational tool. Also, it can be very easily extended to continuous time systems governed by differential equations. Second, the condition (4.7) of minimization of the current value Hamiltonian does not necessarily lead to a difference equation, like the adjoint equations and hence it can be very easily interpreted in economic terms.

Note that the objective function in our example is so chosen that it is strictly concave and differentiable in both the state and control variables. Hence the necessary conditions are also sufficient. In the general case however sufficiency conditions are also needed to characterize the optimal trajectory.

Next we consider a many variable generalization where x is an n-tuple state vector, u(t) an m-tuple control vector and the state dynamics is a system of n linear difference equations

$$x(t+1) = A\,x(t) + B\,u(t)$$
$$x(0) = x_0 \text{ given.} \tag{5.1}$$

The objective function is to minimize J i.e.

$$\text{Min } J = \sum_{t=0}^{N-1} [(1/2)\,x^T(t)\,Q\,x(t) + (1/2)\,u^T(t)\,R\,u(t)] \tag{5.2}$$

where the square matrix R of order m is assumed to be positive definite (symmetric) and the square matrix Q of order n to be positive semidefinite (symmetric). These conditions on matrices R and Q make the objective function J strictly convex in the control variables and convex in the state variables, so that the necessary conditions are also sufficient.

The optimal control problem is to choose the time path or trajectory of the control vector u(t), $0 \le t \le N-1$ so that J in (5.2) is minimized subject to the state dynamic equations (5.1). Note that the terminal period is denoted here by N-1, since the superscript T on x and u is used to denote the transpose of vectors. Denote the Lagrangean and the current value Hamiltonian by L and H(t)

$$L = J + y^T(t+1)\,[A\,x(t) + B\,u(t) - x(t+1)]$$
$$H(t) = (1/2)\,[x^T(t)\,Q\,x(t) + u^T(t)\,R\,u(t)]$$
$$\quad + y^T(t)\,[A\,x(t) + B\,u(t)].$$

Then the optimal trajectory must satisfy the following necessary conditions for t = 0,1,...,N-1

$$\frac{\partial L}{\partial x(t)} = Q\,x(t) + A^T\,y(t+1) - y(t) = 0 \tag{5.3}$$

$$\frac{\partial L}{\partial u(t)} = R\,u(t) + B^T\,y(t+1) = 0 \tag{5.4}$$

$$\frac{\partial L}{\partial y(t+1)} = A\,x(t) + B\,u(t) - x(t+1) = 0 \tag{5.5}$$

and the transversality condition

$$\frac{\partial L}{\partial x(N)} = y(N) = 0. \tag{5.6}$$

The condition (5.3) yields

$$y(t) = Q\, x(t) + A^T y(t+1).\qquad(5.7)$$

Also from (5.4), $\quad u(t) = -R^{-1} B^T y(t+1).\qquad(5.8)$

On substituting these into the system dynamics (5.1)

$$x(t+1) = A\, x(t) - B R^{-1} B^T y(t+1).\qquad(5.9)$$

We now stipulate an important transformation known as the Riccati transformation relating the state vector and the vector of Lagrange multipliers

$$y(t) = P(t)\, x(t)\qquad(5.10)$$

where $P(t)$ is a square matrix of order n. On using this transformation in (5.7) and (5.9) and eliminating $y(t)$ we obtain

$$P(t)\, x(t) = Q\, x(t) + A^T P(t+1)\, x(t+1)\qquad(5.11)$$
$$x(t+1) = A\, x(t) - B R^{-1} B^T P(t+1)\, x(t+1).\qquad(5.12)$$

Solving for $x(t+1)$ in (5.12) yields

$$P(t)\, x(t) = Q\, x(t) + A^T P(t+1)\, M(t)\, A\, x(t)\qquad(5.13)$$

where $\quad M(t) = [I + B R^{-1} B^T P(t+1)]^{-1}.$

Since (5.13) must hold for all $x(t)$ it must satisfy

$$P(t) = Q + A^T P(t+1)\, M(t)\, A.\qquad(5.14)$$

This is a recursive relation for the matrix $P(t)$ used in the Riccati transformation. It must be solved backwards starting with $P(N)$. Since $y(N) = P(N)\, x(N)$ and $y(N) = 0$ we have $P(N) = 0$. With $P(t)$ solved, we can compute the optimal control vector by eliminating $y(t+1)$ from (5.7) and (5.8)

$$\begin{aligned}u(t) &= -R^{-1} B^T (A^T)^{-1} [y(t) - Q\, x(t)] \\ &= -R^{-1} B^T (A^T)^{-1} [P(t) - Q]\, x(t) \\ &= K(t)\, x(t)\end{aligned}\qquad(5.15)$$

where K(t) is of the feedback form

$$K(t) = -R^{-1} B^T (A^T)^{-1} [P(t) - Q].$$

This important property of optimal control related to the state vector by a linear feedback rule, which is comparable to the multiplier-accelerator formulation in Keynesian macro-dynamics holds more generally when the state dynamics has an additive random noise vector v(t)

$$x(t+1) = A\, x(t)\ \ B\, u(t) + v(t)$$

and the objective function is changed to minimization of the expected value of J. If the noise term v(t) has a zero mean Gaussian (normal) distribution with a constant covariance for all t and a distribution independent of the control variables, then this type of optimal control model is called LQG i.e. linear, quadratic and Gaussian.

The LQG models have been most widely applied in control engineering due to the ease of application and the fact that many nonlinear problems can be suitably approximated by the LQG models. Economic applications include large-scale econometric models used as policy models, computable general equilibrium models used as planning models and various models of macro-dynamic policymaking based on monetary and other sectors. For centrally controlled planned economies like Hungary, such control models have been widely used on the basis of dynamic input-output models. However, due to parameter uncertainty these models frequently take on the character of a stochastic control, which does not always satisfy the standard LQG assumptions. Hence many important questions about the sufficiency conditions remain intractable in the realm of optimal stochastic control.

4.2 Intertemporal Optimization: Continuous Time

In continuous time cases the system dynamics is usually represented by ordinary differential equations and the discrete intertemporal sum in the objective function replaced by an integral which is nothing but a continuous sum over a given horizon. Thus the neoclassical model of optimal growth given in (3.2) and (3.3) is transformed as follows

$$\text{Max } J = \int_0^T U(c) \exp(-\delta t) dt \tag{6.1}$$

subject to $\quad Dk(t) = f(k) - (n+r)k - c$

$$k(0) = k_0 \text{ given} \tag{6.2}$$

where $D = d/dt$, $\exp(-\delta t) = e^{-\delta t}$ denotes continuous discounting at the positive rate δ and time is implicit in the state variable $k = k(t)$ and the control variable $c = c(t)$. The decision problem before the social planner is: what should be the time path or trajectory of per capita consumption $c(t)$ which satisfies the system dynamics (6.2) and also maximizes the discounted sum of utilities over the given horizon $[0,T]$? We can solve such problems in two ways. One is the Euler-Lagrange method of variational calculus, the other being the continuous form of Pontryagin's maximum principle. The second one is more general in that it can handle inequality restrictions on the state and control variables.

Assume for simplicity that the utility function $U(c)$ is twice differentiable with positive but diminishing marginal utility for all positive levels of per capita income i.e.

$$dU(c)/dc = U'(c) > 0, \quad dU'(c)/dc < 0, \text{ all } 0 < c < \infty$$

and for meaningful results we also assume

$$\lim_{c \to 0} U'(c) = \infty, \quad \lim_{c \to \infty} U'(c) = 0$$

i.e. consumption is always desirable and it is bounded from above. By using the Lagrange multiplier $p(t)$ we rewrite the problem as one of maximizing the unconstrained functional $\bar{J}$

$$\bar{J} = \int_0^T \exp(-\delta t)[U(c) + p(t)\{f(k) - (n+r)k - c - Dk\}]dt$$

$$= \int_0^T [\exp(-\delta t) U(c) + \hat{p}(t)\{f(k) - (n+r)k - c - Dk\}]dt$$

$$= \int_0^T F(k, Dk, c, \hat{p}) \, dt, \quad \hat{p}(t) = p(t)\exp(-\delta t).$$

The Euler-Lagrange necessary conditions for the maximum of $\bar{J}$ require that

$$\frac{\partial F}{\partial q(t)} = \frac{d}{dt}\left(\frac{\partial F}{\partial (Dq)}\right), \quad Dq = dq(t)/dt \quad (6.3)$$

where $q(t)$ is any of the three variables k, c and $\hat{p}$. These necessary conditions are also sufficient by the strict concavity of $U(c)$. On applying the Euler-Lagrange conditions (6.3) to the integrand $F = \exp(-\delta t)U(c) + \hat{p}(t)\{f(k) - (n+r)k - c - Dk\}$ we obtain

$$Dp(t) = \delta p(t) - f'(k) p(t) + (n+r) p(t) \tag{6.4}$$

$$U'(c) = p(t), \quad \text{where } f'(k) = \partial f(k)/\partial k \tag{6.5}$$
$$U'(c) = \partial U(c)/\partial c.$$

Together with the dynamic equation (6.2) we have three equations to compute the optimal trajectories of the three variables: $k(t)$, $c(t)$ and $p(t)$.

The same equations could be obtained in terms of Pontryagin's principle by defining the current value Hamiltonian as

$$H(t) = \exp(-\delta t) U(c) + \hat{p}(t) \{f(k) - (n+r)k - c\}$$

and applying the necessary conditions in two parts. The first part is dynamic and uses the two adjoint equations

$$Dk = dk/dt = \partial H/\partial \hat{p}(t)$$
$$D\hat{p}(t) = -\partial H/\partial k(t), \qquad \hat{p}(t) = p(t) \exp(-\delta t). \tag{6.6}$$

These yield (6.2) and (6.4). The second part is essentially static, since it holds for each t and it requires that $H(t)$ is maximized with respect to the control variable c at each t. This yields through $\partial H(t)/\partial c = 0$ the condition (6.5) which says that the marginal utility of per capita consumption must equal the dynamic shadow price $p(t)$ along the optimal trajectory for all t. Note that if there are additional constraints on the control variable in the form of inequalities these can be handled by Kuhn-Tucker theory e.g. let there be a constraint $c \geq c_o$, $c_o > 0$ imposed by the planner. By using the Lagrange mutliplier λ we would consider the relevant part of the $H(t)$ function as $\bar{H} = U(c) - pc + \lambda(c - c_o)$ by dropping the terms not required. For optimal c, we require $U'(c) - p + \lambda \leq 0$, $c \geq c_o$ by Kuhn-Tucker theory i.e. if $c > c_o$, then $\lambda = 0$ and $U'(c) = p$ but otherwise $c = c_o$, $\lambda > 0$ and $U'(c) \leq p - \lambda$.

One advantage of the Pontryagin principle is that the steady-state equilibrium values of k, p and c can be easily located from (6.6) by setting $Dk = 0 = D\hat{p}(t)$ and from the static condition $\partial H(t)/\partial c = 0$. We obtain these values k^*, p^*, c^* from the algebraic equations as follows

$$f'(k^*) = n+r+\delta$$
$$f(k^*) = (n+r)k^* + c^* \tag{6.7}$$
$$U'(c^*) = p^*.$$

As before we obtain the Ramsay rule for $n = r = 0 = \delta$ and the Phelps-Swan rule when $r = 0 = \delta$ and $n > 0$. Arrow has interpreted the last two equations of (6.7) as "derived demand functions" for consumption and capital. For example if $U(c) = c - (a/2)c^2$, then the last equation yields $c^* = (1/a)(1 - p^*)$ with $\partial c^*/\partial p^* < 0$ i.e. a derived demand function with a negative slope. This suggests that if we start near the equilibrium point (c^*, k^*, p^*), one may converge again to the same point, provided certain conditions are fulfilled on the characteristic roots. Two ways one can show this. One is to use the two autonomous differential equations in c and k from (6.5) and (6.6) after eliminating p and then linearize around (c^*, p^*) by Taylor series expansion. For this reduced equation system one could check if the two characteristic roots are real and opposite in sign, in which case they define a saddle point equilibrium. A second method is to use a quadratic form for $U(c) = c - (a/2)c^2$ and a linear production function $f(k) = bk$ and then derive from (6.5) and (6.6) the following autonomous linear differential equations i.e. equations with constant coefficients

$$Dk = gk - c, \quad g = b - n - r$$
$$Dc = 0k + hc - (h-1)/a \qquad h = \delta - g. \qquad (6.8)$$

The characteristic equation is $\lambda^2 - (g+h)\lambda + gh = 0$, which has two roots g, $\delta - g$ which are real distinct and of opposite sign if $\delta < g$. Thus if (6.8) describes the paths of $k(t)$, $c(t)$ close to the steady-state equilibrium point (k^*, c^*), which is also called a balanced growth point, then the local stability is analyzed by the differential equation system (6.8). It can be shown that if the initial capital stock k_o is such that $k_o < k^*$, then both $c^* = c^*(t)$ and $k^* = k^*(t)$ will optimally increase over time, moving up the stable branch to the balanced growth equilibrium, whereas if $k_o > k^*$, then both $c^*(t)$, $k^*(t)$ optimally decrease over time, moving down the stable branch to the balanced-growth equilibrium

$$\lim_{t \to \infty} c^*(t) = c^*, \qquad \lim_{t \to \infty} k^*(t) = k^*.$$

Thus the steady-state balanced growth equilibrium point (k^*, c^*) has saddle point stability.

4.2.1 Optimal Model of Advertisement

A class of monopoly price-advertising models has been recently applied in marketing literature, which maximizes the discounted sum of profits

$$\max_{u,p} J = \int_0^T z(t) \exp(-rt) dt \qquad (7.1)$$

where

$$z(t) = \text{profits} = [p(t) - c(x)]Dx(t) - g(u(t))$$

and

$$Dx(t) = dx(t)/dt = f(p,u,x) \tag{7.2}$$
$$x(0) = x_0 \text{ given}$$

where $x = x(t)$ is cumulative output or sales at time t, $u(t)$ = the advertisement rate, $g(u) = ku(t)$ is the linear advertising cost, p = price, $c(x)$ = production cost as a function of cumulative output, $Dx = dx/dt$ and $f(p,u,x)$ is the demand function for which we assume the form

$$f(p,u,x) = a_0 - a_1 x + a_2 u - (a_3/2) u^2 - bp \tag{7.3}$$

where the parameters are assumed positive. Clearly $\partial f/\partial u > 0$ and $\partial^2 f/\partial u^2 < 0$ implying that advertisement has a diminishing marginal impact on the rate of current sales, whereas increasing price tends to reduce the rate of current sales i.e. $\partial f/\partial p < 0$.

Define the current value Hamiltonian $H(t)$

$$H(t) = \exp(-rt)[\{p(t) - c(x)\} f(p,u,x) - g(u)] + \hat{\lambda}(t) f(p,u,x) \tag{7.4}$$

where $\hat{\lambda} = \lambda(t) \exp(-rt)$, $\lambda(t)$ is the adjoint variable (i.e. dynamic shadow price) and r is the positive discount rate. By Pontryagin's maximum principle the optimal trajectories must satisfy the two adjoint equations

$$Dx(t) = \partial H(t)/\partial \hat{\lambda}(t)$$
$$D\hat{\lambda}(t) = -\partial H(t)/\partial x(t) \tag{7.5}$$

and $H(t)$ must be a maximum with respect to the control variables u and p for each t i.e.

$$\partial H(t)/\partial p = 0, \quad \partial H(t)/\partial u = 0 \tag{7.6}$$

and if the terminal period T of the planning horizon is fixed but $x(T)$ is free to vary then we must also satisfy the transversality condition

$$\lambda(T) = 0. \tag{7.7}$$

From (7.5) we obtain

$$D\lambda(t) = r\lambda - (p - c + \lambda)f_x + fc_x$$

where $f_x = \partial f/\partial x$, $c_x = \partial c(x)/\partial x$

and (7.6) yields for the optimal p and optimal u

$$p(t) = c - \lambda - (f/f_p), \quad f_p = \partial f/\partial p = -b$$
$$= (\frac{e}{e-1})[c(x) - \lambda], \quad e = -p\, f_p/f \text{ (price elasticity of demand)}$$
$$u(t) = (a_2/a_3) - (k/a_3)[p - c + \lambda]^{-1}.$$

Several interesting results may be derived by making different assumptions about the price elasticity parameter e.

(i) If the price elasticity is constant, then $Dp(t) = dp(t)/dt = -e(e-1)^{-1}$ $f\, f_x/f_p$, i.e. optimal price increases (decreases) if (e-1) and f_x have the same (opposite) sign.

(ii) If $f_x = 0$, i.e. $a_1 = 0$, then from (i) we get $Dp(t) = 0$ i.e. optimal price is constant i.e. $p(t) = e(e-1)^{-1} c(x(T))$, since $\lambda(T) = 0$ by transversality where the condition $e > 1$ is required for positivity of price.

(iii) The steady state equilibrium values x^*, λ^*, p^*, u^* satisfy $x^* = (f/a_1) - \lambda^*(1 + r/a_1)/c_1 - p^*/c_1$; $u^* = (a_2/a_3) - (k/a_3)(e-1)/(c_0 - c_1 x^* - \lambda^*)$ when $c(x) = c_0 - c_1 x$ shows declining marginal costs i.e. $\partial c(x)/\partial x = -c_1 < 0$. Clearly this shows that optimal cumulative sales (x^*) declines as price p^*, or the adjoint variable λ^* or the discount rate rises. Also the optimal advertisement rate u^* falls as the price elasticity rises.

4.2.2 Continuous Pontryagin Principle

The general version of the continuous Pontryagin princple, when $x(t)$ and $u(t)$ are state and control vectors of orders n and m respectively considers the problem of minimizing J i.e.

$$\text{Min } J = h(x(T),T) + \int_0^T g(x(t), u(t), t)dt \qquad (8.1)$$

subject to

$$Dx_i(t) = dx_i/dt = f_i(x(t), u(t), t) \qquad i=1,2,\ldots,n \qquad (8.2)$$
$$x(0) = x_0 \text{ given}$$

where $h(\cdot)$, $g(\cdot)$ and each $f_i(\cdot)$ are assumed to be scalar convex and differentiable functions of their arguments. Pontryagin principle states that if there exists an optimal trajectory $\{u(t), 0 \leq t \leq T\}$ and the associated time path of $\{x(t), 0 \leq t \leq T\}$, then they must satisfy the following necessary conditions defined in terms of the Hamiltonian H

$$H = H(x,u,\lambda,t) = g(x,u,t) + \sum_{i=1}^{n} \lambda_i f_i(x,u,t)$$

(1) $\quad dx_i/dt = \dfrac{\partial H}{\partial \lambda_i}, \qquad i=1,2,\ldots,n$

(2) $\quad d\lambda_i/dt = -\partial H/\partial x_i, \qquad i=1,2,\ldots,n$

(3) H attains its minimum with respect to each u_i e.g., if H is not linear in u, then $\partial H/\partial u_j = 0$, $j=1,2,\ldots,m$ characterizes the minimum since H is convex and differentiable in $u = (u_j)$,

(4) $\partial h/\partial x_i(T) = \lambda_i(T)$, $i=1,2,\ldots,n$ and $x(0) = x_0$ when the initial state x_0 is fixed and given and the terminal time T is fixed but $x(T)$ may vary. The first condition, also known as the transversality condition reduces to $\lambda_i(T) = 0$, $i=1,2,\ldots,n$ if there is no scrap value function such as $h(\cdot)$ in (8.1).

The above conditions are only necessary for the minimum; we need additional conditions known as the sufficiency conditions. Two of the most common situations where the above necessary conditions are also sufficient are the following: (i) LQG model where $g(\cdot)$ is quadratic, each $f_i(\cdot)$ is linear and $h(\cdot)$ is linear or quadratic and (ii) the case where the function $\bar{H}(x,\lambda,t) = \max H(x,u,\lambda,t)$ is a concave differentiable function of x for given λ and t.

Transversality Conditions

The end point conditions $\partial h/\partial x_i(T) = \lambda_i(T)$, $i=1,2,\ldots,n$ holding at the terminal boundary T need some clarification. In general, when both T and $x(T)$ are not fixed, the transversality condition requires for each $i=1,2,\ldots,n$

$$\sum_{i=1}^{n} \left(\dfrac{\partial h(x(T),T)}{\partial x_i} - \lambda_i(T)\right) \delta x_i(T)$$

$$+ \left[H(x(T), u(T), \lambda(T),T) + \dfrac{\partial h(x(T),T)}{\partial t}\right] \delta T = 0 \qquad (8.3)$$

where $\delta x_i(T)$, δT are the first variations of $x_i(T)$ and T along the optimal trajectory.

(i) If the terminal time T is fixed, we have two cases: $x(T)$ fixed and $x(T)$ free. If T and $x(T)$ are both fixed, then $x(T) = x_T$ must be given, since $\delta x(T) = 0 = \delta T$. If T is fixed but the final state $x(T)$ is free to vary, then $\delta T = 0$ in (8.3) and hence $\partial h(\cdot)/\partial x_i(T) = \lambda_i(T)$.

(ii) If the terminal time T is free to vary, we have again two cases: $x(T)$ fixed or free to vary. In the first case, we have $\delta x(T) = 0$ in (8.3) which imply the transversality condition

$$H(x(T), u(T), \lambda(T), T) + \frac{\partial h(x(T),T)}{\partial t} = 0.$$

On the other hand if the final state is free, then $\delta x(T)$, δT are both arbitrary and independent so that their coefficients are zero i.e. we obtain the n equations

$$\lambda_i(T) = \frac{\partial h(x(T),T)}{\partial x_i} \tag{8.4}$$

and the additional equation

$$H(x(T), u(T), \lambda(T),T) + \frac{\partial h(x(T),T)}{\partial t} = 0 \tag{8.5}$$

specifying the terminal conditions to be satisfied by the optimal trajectory.

Thus it is clear that in the neoclassical model (6.1), (6.2) if T were infinity (i.e free), then we would require the satisfaction of (8.4) and (8.5) as the transversality conditions i.e.

$$\lim_{T \to \infty} \exp(-\delta T) \, p(T) = 0$$

and

$$\lim_{T \to \infty} H(T) = 0, \; H(T) = \exp(-\delta T)[U(c) + p(T) \{f(k) - (n+r) \, k-c\}]$$

LQG Model

As in the discrete time case, we may now consider a deterministic version of the LQG model by ignoring the additive error term ($v = 0$) in the system dynamics

$$Dx(t) = Ax + Bu + v \tag{9.1}$$
$$x(0) = x_0.$$

The objective function is

$$\text{Min } J = (1/2) \int_0^T [x^T Q x + u' R u]dt \tag{9.2}$$

where R is assumed to be symmetric positive definite, Q symmetric and positive semidefinite and A, B are suitable constant matrices. The terminal time T is assumed fixed but x(T) may vary. Hence we need the transversality condition $\lambda(T) = 0$, where $\lambda(t)$ is the adjoint vector (i.e. dynamic shadow price vector) in the current Hamiltonian $H(t) = (1/2)[x^T Q x + u' R u] + \lambda^T (A x + B u)$. By applying the two necessary conditions (which are also sufficient here) of the Pontryagin principle i.e.

$Dx(t) = \partial H/\partial \lambda(t)$ i.e. $\dfrac{dx_i(t)}{dt} = \dfrac{\partial H}{\partial \lambda_i(t)}$, $i=1,2,\ldots,n$

$D\lambda(t) = -\partial H/\partial x(t)$ i.e. $\dfrac{d\lambda_i(t)}{dt} = \dfrac{\partial H}{\partial x_i(t)}$, $i=1,2,\ldots,n$

$\partial H(t)/\partial u = 0$ i.e. $\dfrac{\partial H(T)}{\partial u_j(t)} = 0$, $j=1,2,\ldots,m$.

One gets

$$u = -R^{-1} B^T \lambda$$
$$Dx = Ax - BR^{-1}B^T \lambda \qquad (9.3)$$
$$D\lambda = -Qx - A^T \lambda.$$

One method of solving the above two linear differential equation systems in x and λ is to use the Riccati transformation $\lambda = Px$, where the subscript t is omitted from λ, x, u and the matrices P, R, Q, B, A. The use of Riccati transformation in (9.3) where $P = P(t)$ is a square matrix of order n yields a matrix differential equation in P otherwise known as the Riccati equation

$$DP = \dfrac{dP}{dt} = -PA - A^T P + PBR^{-1} B^T P - Q. \qquad (9.4)$$

We solve this equation for P under the boundary condition $P(T) = $ zero matrix. On using this optimal P we obtain the optimal control vector u as a linear feedback rule

$$u = K(t) x(t), \quad K(t) = -R^{-1}(t) B^T(t) P(t). \qquad (9.5)$$

In order to obtain the optimal trajectory, the matrix Riccati equation (9.4) is integrated backwards from T with $P(T) = 0$ to $t=0$. The initial condition $P(0) = P_o$ is then reached. Then we integrate the differential equations $Dx = Mx$, $M = A - BR^{-1} B^T$ P forward with the initial conditions $x(0) = x_o$, $P(0) = P_o$. Computer routines are available for solving the Riccati equation (9.4) and hence the LQG models.

The LQG models have been widely applied in specifying optimal monetary and fiscal policies under a linear macro-dynamic Keynesian model, where the objective function usually takes the form of minimizing a weighted sum of deviations from some desired or target values $\bar{x}_i$, $\bar{u}_j$ of the policymaker e.g.

$$\text{Min } J = \int_0^T [\sum_{i=1}^n \alpha_i (x_i - \bar{x}_i)^2 + \sum_{j=1}^m \beta_j (u_j - \bar{u}_j)^2] dt$$

with α_i, β_j as nonnegative weights.

One major reason for applying the LQG models so widely is the presence of the additive error term v in (9.1) when it is normally distributed with a fixed variance-covariance matrix. In this case a result known as the certainty equivalence theorem holds, which implies among other things that the optimal linear feedback rule (9.5) can be successively used to update the optimal control as more and more information becomes available to provide revised estimates of the state vector.

As a scalar example of the LQG model consider a linear Keynesian model

$$Dx = ax + bu, \quad D = d/dt$$
$$x(0) = x_0 \text{ given}$$

where x is national income, u is government expenditure and the policymaker's objective is to

$$\text{Min } J = (1/2) \int_0^T [(x - \bar{x})^2 + (u - \bar{u})^2] dt$$

where the desired values $\bar{x}$, $\bar{u}$ are set to zero by suitable choice of units. By applying Pontryagin principle we get for $H = \tfrac{1}{2}(x^2 + u^2) + \lambda(ax + bu)$

$$d\lambda/dt = -\frac{\partial H}{\partial x} = -(x + \lambda a), \quad \frac{dx}{dt} = \frac{\partial H}{\partial \lambda} = ax + bu$$

$$\frac{\partial H}{\partial u} = 0 = u + \lambda b.$$

Hence

$$Dx = ax - b^2 \lambda$$
$$D\lambda = -x - a\lambda \qquad (9.6)$$

The characteristic equation in terms of the root s is

$$s^2 - (a^2 + b^2) = 0$$

i.e.

$$s_1, s_2 = \pm \sqrt{(a^2 + b^2)}.$$

The roots are real and opposite in sign i.e. they have the saddle point property. Hence the optimal state trajectory is

$$x(t) = A_1 \exp(s_1 t) + A_2 \exp(s_2 t). \qquad (9.7)$$

By using this and $Dx(t)$ in the first equation of (9.6) we get

$$\lambda(t) = (a - s_1)(A_1/b^2) \exp(s_1 t) + (a - s_2)(A_2/b^2) \exp(s_2 t)$$

but $\lambda(T) = 0$ by transversality, hence

$$0 = (a - s_1)(A_1/b^2) \exp(s_1 T) + (a - s_2)(A_2/b^2) \exp(s_2 T)$$

also from (9.7) with $x(0) = x_0$ at $t = 0$ we get

$$x_0 = A_1 + A_2.$$

Hence A_1 and A_2 can be solved as

$$\begin{matrix} A_1 \\ A_2 \end{matrix} = \begin{matrix} g_1(T) & g_2(T) \\ 1 & 1 \end{matrix}^{-1} \begin{matrix} 0 \\ x_0 \end{matrix}$$

where

$$g_1(T) = (a - s_1) \exp(s_1 T)/b^2$$
$$g_2(T) = (a - s_2) \exp(s_2 T)/b^2.$$

Note that if T were infinity, then the optimal trajectory $x(t)$ in (9.7) would be infinity as $t \to \infty$ but this is impossible, since $u(t)$ would also be unbounded. Hence A_1 should be zero in (9.7) because s_1 is the positive root leading to unboundedness of the optimal trajectory.

4.2.3 Economic Applications of Optimal Control

We consider several examples of optimal control to illustrate the typical economic problems which arise in applications e.g. the optimal control trajectory may be piecewise continuous, there may be zones of stability and instability in the control-cum-output space and there may be stochastic elements present in the control law.

Ex 1. (Longrun goodwill problem)

Consider the net return $z = R(x) - C(u)$ of a firm as the difference between total revenue $R = R(x)$ depending on goodwill state x and total cost $C(u)$ depending on advertisement $u = u(t)$. The dynamics of the goodwill state is

$$Dx = dx/dt = -mx + nu \qquad (10.1)$$
$$x(0) = x_0 \text{ given.}$$

The firm's objective is to

$$\text{Maximize } J = \int_0^\infty \exp(-rt)\, z(t)\, dt \qquad (10.2)$$

subject to (10.1), where m,n are fixed positive parameters. It is assumed that $R(x)$ is a twice diffrentiable strictly concave function and $C(u)$ is a nonlinear convex cost function with $\partial C(u)/\partial u > 0$ and $\partial^2 C(u)/\partial u^2 > 0$. The Hamiltonian is

$$H = \exp(-rt)[R(x) - C(u) + p(nu - mx)]$$

with $p = p(t)$ as the adjoint variable. Applying the necessary conditions of the Pontryagin maximum principle, which are also sufficient since $z(t)$ is strictly concave in x and u, we obtain

$$dp/dt = (r+m)p - \partial R/\partial x$$
$$p = (\partial C/\partial u)/n$$
$$dx/dt = nu - mx, \qquad x(0) = x_o.$$

Due to the discounted infinite time problem, the transversality condition is

$$\lim_{T \to \infty} p(T) \exp(-rT) = 0$$

which implies $\lim_{T \to \infty} (\partial C/\partial u(T)) \exp(-rT) = 0$

i.e. this requires the boundedness of marginal cost $\partial C/\partial u$. Using the simplification n=1 (if necessary by a change of units of u) and taking the derivative of $C' = \partial C/\partial u$, we get a system of simultaneous equations in x and u

$$du/dt = (1/C'') \{(r+m) C' - R'\};$$
where $\quad C' = \partial C/\partial u, \; C'' = \partial C'/\partial u \qquad (10.3)$
$$dx/dt = u - mx.$$

These equations specifying the optimal trajectory can be investigated by phase diagrams. To construct such a diagram we assume a saddle point defined by $Dx = 0$ and $Du = 0$ and check the singular trajectories leading to this point. Specifying for the singular curve $Dx = 0$ we get $u = mx$, but the singular curve for $du/dt = 0$ yields

$$(r+m) C'(u) = R'(x)$$

so that differentiating both sides of this with respect to x yields

$$(r+m) C'' \frac{du}{dx} = R''(x)$$

or,

$$\left(\frac{du}{dx}\right)_{Du=0} = \frac{R''(x)}{(r+m)C''} \leq 0, \text{ since } R''(x) < 0$$

thus the singular curve for Du = 0 has a negative slope. Now at points above Dx = dx/dt = 0 we have Dx > 0 and below this curve, Dx < 0. Furthermore since

$$\partial(Du)/\partial x = -R''/C'' > 0$$

it follows that at points to the right of the Du = 0 curve, Du is positive, while for points to the left of Du = 0, Du is negative. Thus, if the equilibrium point (x*, u*) defined by Du = 0, Dx = 0 is a saddle point, then there are exactly two trajectories of the system of differential equations which lead to this equilibrium as t→∞. Linearizing the system (10.3) around the stationary point (x*, u*) we obtain

$$\begin{pmatrix} Dx \\ Du \end{pmatrix} = \begin{bmatrix} \alpha_{11} & \alpha_{12} \\ \alpha_{21} & \alpha_{22} \end{bmatrix} \begin{pmatrix} x \\ u \end{pmatrix} \quad (10.4)$$

where the α_{ij} are the partial derivatives of the system equations dx/dt, du/dt in (10.3). These are evaluated at the point x*, u*, making both Dx and Du equal to zero simultaneously. The eigenvalues of the coefficient matrix (α_{ij}) in (10.4) are given by the two characteristic roots

$$\lambda_1, \lambda_2 = (1/2) [r \pm \{r^2 + m(r + m) - \frac{R''}{C''}\}^{\frac{1}{2}}]$$

which are real and opposite in sign since the term m(r+m) - R''/C'' is positive. Hence it defines a saddle point. Thus this unique path to (x*, u*) is optimal i.e. all other paths lead either to infinitely large x and u or, to a zero level of goodwill.

<u>Ex 2</u>. (Bang-bang control)

Consider a macro-dynamic Keynesian model

$$Y = C + I + G, \quad C = cY, \quad I = v \, dY/dt$$

where G is government expenditure, the control variable. The reduced form equation is DY = aY + bG, a = (1-c)/v, b = -1/v. Change the notation as Y = x = x(t), G = u

$$\text{maximize } J = \int_0^T (x - u)\,dt$$

subject to

$$Dx = ax + bu, \qquad A \leq u \leq B$$
$$x(0) = x_0 \text{ given}$$

where A, B and x_0 are given constants and the length T of the horizon is fixed. According to Pontryagin's maximum principle, it is necessary that an optimal control path $u = u(t)$, $0 \leq t \leq T$ be chosen such that the Hamiltonian $H = x - u + p(ax + bu)$ is maximal, where p is the adjoint variable. But since H is linear in u, the optimal value of u must be in three parts:

$$u = u_{opt} = \begin{cases} B, & \text{if } -1 + pb > 0 \\ \text{undefined}, & \text{if } -1 + pb = 0 \\ A, & \text{if } -1 + pb < 0 \end{cases}$$

When $-1 + pb = 0$ holds on a finite number of points, the optimal control law is said to be of the bang-bang type, since it switches between A and B.

Ex 3. (Two-sector linear growth model)

This two-sector neoclassical growth model, due to Michael Bruno (1967) combines the linear programming framework of activity analysis in production with the optimal neoclassical growth. Solow's growth equation $Dk = f(k) - c - (n+r)k$ is written as $Dk = z - \lambda k$, $z = f(k) - c$ = per capita investment good, $\lambda = n+r$ and the optimal policy is to choose the two control variables c (consumption good) and z (investment good) in per capita terms so as to

$$\text{maximize } J = \int_0^\infty c \exp(-\delta t)\,dt \tag{11.1}$$

subject to

(A) $\quad \begin{aligned} a_0 c + a_{01} z &\leq 1 \text{ (labor constraint)} \\ a_1 c + a_{11} z &\leq k \text{ (capital constraint)} \\ c, z &\geq 0 \text{ (nonnegativity)} \end{aligned}$ \hfill (11.2)

and

(B) $\quad \begin{aligned} Dk &= dk/dt = z - \lambda k \\ k(0) &= k_0 > 0 \text{ given.} \end{aligned}$ \hfill (11.3)

Here $k = k(t)$ is the state variable, $c = c(t)$, $z = z(t)$ are the two control variables, δ is the fixed positive rate of discount and all other variables are positive scalars. The linear constraints in (A) are discrete in time and static in some sense because they do not involve differential equations, whereas those in (B) are dynamic. For the dynamic part we form the Hamiltonian

$$H(t) = \exp(-\delta t) [c + p(z - \lambda k)] \qquad (11.4)$$

and state that if there exists an optimal trajectory or program $\{c(t), z(t), k(t); 0 \le t \le \infty\}$, then there must exist a continuous function $p = p(t)$ of the dynamic shadow price such that

$$Dk = -\lambda k + z \text{ with } k(0) = k_o \text{ (admissibility)}$$
and $Dp = dp/dt = (\lambda + \delta) p - s$, (perfect foresight condition) $\qquad (11.5)$

and at each moment of time $H(t)$ is maximized with respect to the two control variables subject to the constraints in (A). This yields a linear programming (LP) problem whose dual is

$$\begin{aligned}
\text{minimize } & C = w + sk \\
\text{s.t. } & w\, a_o + s\, a_1 \ge 1 \\
& w\, a_{o1} + s\, a_{11} \ge p \\
& w, s \ge 0
\end{aligned} \qquad (11.6)$$

where w has the interpretation of the real wage (i.e. shadow price of labor) and s that of the rental price of capital (i.e. shadow price of capital).

Also we must have the transversality condition satisfied i.e.

$$\lim_{t \to \infty} p(t) \exp(-\delta t) = 0 \qquad (11.7)$$

and also $p(t)$ needs to be at least piecewise continuous (since it occurs in the LP problem (11.6) above), therefore the optimal trajectory must satisfy the "jump condition"

$$\lim_{h \to 0} [p(\tau + h) - p(\tau - h)] = 0 \qquad (11.8)$$

for the path crossing a phase k boundary at $t = \tau$.

Various regions of the phase space may now be characterized. Suppose we start with insufficient capital stock to ensure full employment of available labor i.e. $k_o < a_{11}/a_{o1}$ and $p \ge a_{11}/a_1$. Then we get from (11.2), $z = k/a_{11}$, $c = 0$ and

$$\frac{Dk}{k} = \frac{1}{a_{11}} - \lambda; \quad \frac{Dp}{p} = \lambda + \delta - \frac{1}{a_{11}}$$

with solutions for $t \leq t_1$:

$$k = k(t) = k_0 \exp(\frac{1}{a_{11}} - \lambda)t$$

$$p = p(t) = A_1 \exp[-(\frac{1}{a_{11}} - \lambda - \delta)t]$$

where the constant A_1 is to be determined later. At time $t = t_1$ we get a corner or "jump" point determined by

$$k_0 \exp[(\frac{1}{a_{11}} - \lambda)t_1] = a_{11}/a_{01} \tag{11.9}$$

and we pass into a new region of full employment where $a_0 c + a_{01} z = 1$, where

$$Dk/k = \frac{a_1}{kG} - \frac{a_0}{G} - \lambda, \quad Dk = 0 \rightarrow k = k^*$$

$$Dp/p = -\frac{a_{01}}{pG} + \frac{a_0}{G} + \lambda + \delta, \quad Dp = 0 \rightarrow p = \hat{p}$$

$$\tag{11.10}$$

where $G = - \begin{vmatrix} a_0 & a_{01} \\ a_1 & a_{11} \end{vmatrix} = \begin{vmatrix} a_{01} & a_0 \\ a_{11} & a_1 \end{vmatrix} = a_{01}a_1 - a_0 a_{11}.$

The solutions of (11.10) are

$$k(t) = (\frac{a_{11}}{a_{01}} - k^*) \exp[-(\lambda + \frac{a_0}{G})(t - t_1)] + k^*$$

$$p(t) = A_2 \exp[(\frac{a_0}{G} + \lambda + \delta)t] + \hat{p}, \quad \text{for } t_2 \leq t < \infty$$

$$\tag{11.11}$$

where A_2 is another constant to be determined. The transversality condition (11.7) requires that A_2 is zero, since $a_0/G + \lambda + \delta$ is positive for a positive G (note that a positive G means that the consumption good is more capital intensive than the investment good). Hence $p(t) = \hat{p}$ for $t_1 \leq t < \infty$. Now we determine A_1 backwards from the jump condition at $t = t_1$

$$\hat{p} = A_1 \exp[-t_1 (\frac{1}{a_{11}} - \lambda - \delta)]$$

and thus for $t \leq t_1$ we must have

$$p(t) = p = \hat{p} \exp[(\frac{1}{a_{11}} - \lambda - \delta)(t_1 - t)]$$

where t_1 is determined from (11.9) as

$$t_1 = (\text{antilog } (\frac{a_{11}}{k_0 a_{01}}))(\frac{1}{1/a_{11} - \lambda})$$

It is clear then that if we start from a stock of initial capital k_0 such that $k_0 > a_1/a_0$, then we decumulate capital by producing c goods only until full employment is reached with $k = a_1/a_0$ and then proceed to the saddle point $(k^*, \hat{p})$.

By using the solution for k one can determine the optimal trajectory for c i.e. in case $k_0 < a_{11}/a_{01}$, then

for $t \leq t_1$, $c = 0$
$t \geq t_1$, $c = (a_{01}k - a_{11})/G$
and hence

$$c = c^* \{1 - \exp[-(\lambda + \frac{a_0}{G})(t - t_1)]\}$$

where $c^* = (a_{01}k^* - a_{11})/G = (1 - a_{11}\lambda)/(a_0 + \lambda G)$.

In case $G < 0$ (i.e. consumption good is less capital intensive than the investment good), then it can be shown that the system reaches full employment only at the optimal point $k = k^*$. Thus the optimal policy for $k_0 \leq k^* \leq a_{11}/a_{01}$ is to accumulate capital and produce no consumption goods until the full employemnt point $k = k^*$ is reached, when at that point $t = t_2$ determined by $k^* = k_0 \exp[t_2 (1/a_{11} - \lambda)]$ we settle at the steady state at (k^*, c^*).

Several interesting features of this model must be noted. First, by using the dual LP model (11.6) and assuming equalities to hold in the first two constraints we can derive the optimal factor price frontier i.e. $s = (1 - a_0 w)/a_1$. Secondly, if $s = 0$, the adjoint equation in p changes from $Dp = (\lambda + \delta)p - s$ to $Dp = (\lambda + \delta)p$. Thus the differential equations can change by the changing values of the dual shadow prices of the static LP model. Lastly, it is clear from (11.11) that for G positive (negative), the steady state k^* of capital is stable (unstable) but the steady state of shadow price $\hat{p}$ is unstable (stable) and hence A_2 has to be set equal to zero. This is sometimes called dual stability-instability property, which is known to hold for the dynamic Leontief IO model and its dual.

Ex. 4 (Adaptive stochastic control)

Consider an investor with wealth x_N at the beginning of period N, part of which is reinvested in a riskless asset with a fixed known return a and the other part reinvested in securities with random returns $b_i = r_i - a$ for security $i=1,2,\ldots,n$. The system dynamics is then

$$x(N+1) = ax(N) + \sum_{i=1}^{n} b_i u_i + e(N+1)$$

$$= ax(N) + b^T u + e(N+1) \qquad (12.1)$$

where the row vector $b^T = (b_1, b_2, \ldots, b_n)$ is random, $u^T = (u_1, u_2, \ldots, u_n)$ is the allocation vector or control and $e(N+1)$ is the stochastic disturbance term assumed to be independently distributed with mean zero and constant variance v for all N. For simplicity it is assumed that a and b are not time dependent. As a simple case we first assume the investor has the short run objective function

$$\min_{u(N)} E(x(N+1) - x^o)^2 \qquad (12.2)$$

where x^o is the target level or goal and the expected value of squared deviations from this target level is minimized by the control vector $u(N)$. The optimal solution is clearly given by

$$[V(N) + b(N) b^T(N)] u(N) = (x^o - ax(N)) b(N) \qquad (12.3)$$

where $b(N) = E\{b|x(N)\}$, $V(N) = \text{var}\{b|x(N)\}$ are the conditional expectation and conditional variance-covariance matrix of vector b, given the observations $x(N)$ on the state. This optimal linear rule (12.3) may be compared with another rule, sometimes called the certainty equivalence (CE) rule when we replace the system dynamics (12.1) by its mean level:

$$x(N+1) = ax(N) + b^T(N) u(N)$$

and the objective function (12.2) by $\min(x(N+1) - \bar{x}^o)^2$. The optimal control law now becomes

$$[b(N) b^T(N)] u(N) = (x^o - ax(N)) b(N). \qquad (12.4)$$

Note two basic differences of this optimal rule (12.4) from that in (12.3). One is that the CE rule (12.4) is not cautious, since the variance term $V(N)$ is absent. Hence this leads to larger variance of control and hence of the objective function. Also this rule does not define the control vector $u(N)$ uniquely, since the term $b(N) b^T(N)$ is not of full rank, so that it cannot be inverted.

Now consider the intertemporal problem with a slightly modified objective function

$$J = E[(1/2) \sum_{N=0}^{M-1} (x(N) - x^o)^2 + s(1 - u^T(N) u(N))]$$

where s is a positive scalar which can be interpreted as a Lagrange multiplier associated with the constraint $\sum_{i=1}^{n} u_i^2(N) = 1$ for all N. By defining a new objective function as $\hat{J}$

$$\hat{J} = J + E\{\lambda(N+1)(a\,x(N) + b^T(N)\,u(N) + e(N+1) - x(N+1))\}$$

we obtain the necessary conditions for the optimal trajectories

$$E\{x(N) - x^o + a\,\lambda(N+1) - \lambda(N)\} = \bar{x}(N) - x^o + a\,\bar{\lambda}(N+1) - \bar{\lambda}(N)$$

and (12.5)

$$E\{s\,u(N) + \lambda(N+1)\,b\} = 0$$

where bar over a variable denotes its expected value. These equations (12.5) have to be recursively solved for the optimal trajectories, subject to the initial condition $x(0) = x_o$ and the transversality condition $\bar{\lambda}(M) = 0$. One approximate way (which is of course only suboptimal) to derive an explicit solution is to assume that the two random terms $\lambda(N+1)$ and b in (12.5) are approximately independent statistically. Then we obtain the optimal rule

$$u(N) = -\bar{\lambda}(N+1)\,(Eb)/s.$$

On using this in the system dynamics and a little algebraic calculation we obtain a set of linear difference equations for N=0,1,2,...,M-1

$$\bar{\lambda}(N+1) = (\bar{\lambda}(N) - \bar{x}(N) + x^o)/a$$
$$\bar{x}(N+1) = k\,\bar{\lambda}(N) + (a - k)\,\bar{x}(N) + k x^o$$
$$\bar{\lambda}(M) = 0, \; x_o \text{ given}, \; k = (as)^{-1}\,(Eb)^T\,(Eb).$$

These equations can be conveniently solved recursively by using the Riccati transformation $\bar{\lambda}(N) = P(N)\,\bar{x}(N)$ and thereby obtaining a closed-loop feedback optimal rule relating u(N) linearly to $\bar{x}(N)$.

4.2.4 Dynamic Programming: Applications

Dynamic programming (DP) provides an alternative algorithm for solving intertemporal optimization problems, whether they are of discrete or continuous time variety. Hence in specific problems like LQG models, where otpimal controls are explicitly computable, the DP algorithm can be very closely related to the Euler-Lagrange conditions of variational calculus and the Pontryagin principle. The DP algorithm applies the optimality principle of Bellman which says that an optimal policy $(u(1),u(2),\ldots,u(T))$ has the property that whatever the initial state $x(0)$ and the initial decision $u(1)$ are, the remaining decisions $(u(2),u(3),\ldots,u(T))$ must constitute an optimal policy for the $(T-1)$-stage process starting in state $x(1)$ which results from the first decision $u(1)$. In other words by Bellman's principle of optimality, an optimal policy has the property that whatever the initial conditions are, the remaining decisions must constitute an optimal policy with regard to the state resulting from the first decision. This principle appears almost as a truism and a proof by contradiction is immediate. Despite its simplicity it provides a powerful algorithm for solving an otherwise intractable problem. To illustrate the principle consider a scalar state variable $x(t)$ and a scalar decision (or control) variable $u(t)$ at time t. Once the initial decision $u(1)$ is made at time $t = t_1$, a new state $x(2)$ results as

$$x(2) = F_1(x(1), u(1))$$

where F is a transformation e.g. it may be a linear function of the arguments. Assume that at this point a second decision is made, resulting in a new state $x(3)$ determined by $x(3) = F(x(2),u(2))$. In general the state $x(n)$ is given by

$$x(n) = F_{n-1}(x(n-1), u(n-1))$$

This is the system dynamics in a forward form. In a backward form this will appear as

$$x(j-1) = F_j(x(j), u(j)) \qquad j=1,2,\ldots,n$$

when there is an n-stage decision problem, each __stage__ containing a number of __states__ e.g. these states are the various possible conditions in which the system may be at a point in time. The states may be finite or infinite for any stage of the decision process and they provide the information required for analyzing the results that the current decision has upon future courses of action, when the future may be looked at usually in a backward form. Let $r_j(x(j),u(j))$ be the return function at stage j and assume that we have the following n-stage maximization problem

$$\underset{u(1),u(2),\ldots,u(n)}{\text{Max}} \quad J = [r_n(x(n),u(n)) + r_{n-1}(x(n-1),u(n-1)) + \ldots + r_1(x(1),u(1))] \qquad (12.1)$$

subject to

$$x(j-1) = F_j(x(j),u(j)) \qquad j=1,2,\ldots,n$$

Note that the objective function is the sum of the individual stage returns. Now define

$$f_n(x(n)) = \max_{u(1),\ldots,u(n)} [r_n(x(n),u(n)) + \ldots + r_1(x(1),u(1))] \qquad (12.2)$$

subject to

$$x(j-1) = F_j(x(j),u(j)) \qquad j=1,2,\ldots,n.$$

Then we obtain

$$f_n(x(n)) = \{\max_{u(n)} \max_{u(n-1),\ldots,u(1)} [r_n(x(n),u(n)) + \ldots + r_1(x(1),u(1))]$$

subject to $\quad x(j-1) = F_j(x(j), u(j)).$

But $r_n(x(n),u(n))$ does not depend on $u(n-1),\ldots,u(1)$. Hence the maximization with respect to these variables can be performed as

$$f_n(x(n)) = \max_{u(n)} \{r_n(x(n),u(n)) + \max_{u(n-1),\ldots,u(1)} [r_{n-1}(x(n-1),u(n-1))$$

$$+ \ldots + r_1(x(1),u(1))]\}.$$

On using (12.2), this can be written as

$$f_n(x(n)) = \max_{u(n)} [r_n(x(n),u(n)) + f_{n-1}(x(n-1))] \qquad (12.3)$$

where

$$x(j-1) = F_j(x(j),u(j)) \qquad j=1,2,\ldots,n.$$

By following the recursion equation one could obtain the optimal solution.

Consider a simple example where we denote $x(j)$ by x_j and $u(j)$ by u_j. The model is

$$\max_{u_j \geq 0} J = \sum_{j=1}^{N} u_j^2$$

subject to
$$\begin{aligned} x_{j-1} &= x_j - u_j, & j=1,2,\ldots,N \\ x_j &\geq 0, & j=1,2,\ldots,N \\ x_n &= c \text{ (fixed)}. \end{aligned} \qquad (12.4)$$

If $j=1$, then by our notation $f_1(x_1) = \max u_1^2$ where $0 \leq u_1 \leq x_1$, since by non-negativity of u_j and x_j we must have $0 \leq u_j \leq x_j$. The maximum is clearly obtained at $u_1 = x_1$, hence

$$f_1(x_1) = \max_{0 \le u_1 \le x_1} u_1^2 = x_1^2.$$

Now we turn to find $f_2(x_2)$ where

$$f_2(x_2) = \max_{0 \le u_2 \le x_2} [u_2^2 + f_1(x_1)] \qquad (12.5)$$

subject to $\qquad x_1 = x_2 - u_2.$ \qquad (12.6)

But (12.5) can be written as

$$f_2(x_2) = \max_{0 \le u_2 \le x_2} [u_2^2 + (x_2 - u_2)^2]$$

by using (12.6). By inspection the optimal solution is at the extreme point $u_2 = x_2$ when

$$f_2(x_2) = x_2^2 \text{ when optimal } u_2 = x_2.$$

By a similar fashion one can show from

$$f_3(x_3) = \max_{0 \le u_3 \le x_3} [u_3^2 + (x_3 - u_3)^2]$$

that the optimal $u_3 = x_3$ and hence

$$f_3(x_3) = x_3^2.$$

Now to use an induction proof assume that

$$f_{n-1}(x_{n-1}) = x_{n-1}^2.$$

Then

$$f_n(x_n) = \max_{0 \le u_n \le x_n} [u_n^2 + f_{n-1}(x_{n-1})] \qquad (12.7)$$

subject to $\qquad x_{n-1} = x_n - u_n.$ \qquad (12.8)

Incorporating the stage transformation function (12.8) in (12.7) we get

$$f_n(x_n) = \max_{0 \le u_n \le x_n} [u_n^2 + (x_n - u_n)^2].$$

The constraints $u_n \le x_n$, $u_n \ge 0$ can be written by adding two nonnegative slack variables v_1^2, v_2^2 as $u_n + v_1^2 = x_n$, $-u_n + v_2^2 = 0$. We denote the Lagrangean function by $L = L(x_n, \lambda_1, \lambda_2, v_1, v_2)$

$$L = u_n^2 + (x_n - u_n)^2 + \lambda_1 (x_n - u_n - v_1^2) + \lambda_2 (u_n - v_2^2).$$

Applying only the necessary conditions of the Kuhn-Tucker Theorem, which of course are not sufficient because the objective function is strictly convex and not concave, we obtain

$$\partial L/\partial u_n = 2u_n - 2(x_n - u_n) - \lambda_1 + \lambda_2 = 0$$
$$\partial L/\partial \lambda_1 = x_n - u_n - v_1^2 = 0$$
$$\partial L/\partial \lambda_2 = u_n - v_2^2 = 0, \quad \partial L/\partial v_1 = -2\lambda_1 v_1 = 0,$$
$$\partial L/\partial v_2 = -2\lambda_2 v_2 = 0 \text{ and}$$
$$\lambda_1, \lambda_2 \geq 0.$$

There are three feasible solutions:

(i) $u_n = x_n$, (ii) $u_n = 0$ and (iii) $u_n = x_n/2$.

Evaluating these we find u_n to be either zero or x_n. Hence

$$f_n(x_n) = x_n^2.$$

Thus $f_N(x_N) = c^2$
$$u_j = c \, \delta_{jk} \qquad j=1,2,\ldots,N \qquad (12.9)$$

where

$$\delta_{jk} = \begin{matrix} 0, j \neq k \\ 1, j = k \end{matrix}$$

This optimal result (12.9) holds for any k=1,2,...,N.

Unfortunately, not all multistage decision problems can be approached or solved by the DP algorithm. Two conditions must be met before Bellman's optimality principle can be invoked: (1) separability of the objective function either additively or multiplicatively, and (2) the state separation property. The first property means that for all k, the effect of the final k stages on the objective function of an n-stage process depends only on state x(n-k) and upon the final k decisions u(n-k+1),u(n-k+2),...,u(n). The state separation property means that after decision u(t+1) is made in stage t+1, the state x(t+1) that results from that decision depends only on x(t) and u(t+1) and does not depend on the previous states x(0),x(1),...,x(t-1). This is also called the Markov property i.e. the dynamic system is memoryless in the sense that the next state in any stage depends only on the current state and current decision.

LQG Model

We now consider an application of dynamic programming to the deterministic version of the LQG model in discrete time form. Let X_t and U_t be n-tuple and m-tuple vectors of state and control variables at time t=1,2,...,T where the horizon T is fixed. The objective is to minimize an intertemporal quadratic loss function (W)

$$W = \sum_{t=1}^{N} (X_t - a_t)^T M_t (X_t - a_t) \tag{13.1}$$

subject to

$$X_t = A_t X_{t-1} + B_t U_t + b_t \qquad t=1,2,\ldots,N \tag{13.2}$$

X_0 given, T is transpose

where a_t is the target vector of desired goals fixed and known, M_t is a known symmetric positive definite matrix and A_t, B_t are known coefficient matrices which may depend on time t. The terminal time N is finite and fixed.

To solve by the DP algorithm one starts at the last period N and move backwards by the recurrence relation. On using all information upto the end of N-1, the objective function (13.1) is written as

$$W_N = X_N^T H_N X_N - 2 X_N^T h_N + d_N \tag{13.3}$$

where for ease of generalization we use

$$H_N = M_N, \quad h_N = M_N a_N, \quad d_N = a_N^T M_N a_N$$

On substituting (13.2) in (13.3) we obtain

$$W_N = (A_N X_{N-1} + B_N U_N + b_N)^T H_N (A_N X_{N-1} + B_N U_N + b_N)$$
$$- 2(A_N X_{N-1} + B_N U_N + b_N)^T h_N + d_N.$$

Then $\partial W_N / \partial U_N = 0$ yields the optimal control denoted for convenience by U_N^*

$$U_N^* = G_N X_{N-1} + g_N \tag{13.4}$$

where

$$G_N = -(B_N^T H_N B_N)^{-1} B_N H_N A_N$$
$$g_N = (B_N^T H_N B_N)^{-1} B_N (h_N - H_N b_N).$$

At this value of U_N^*, the minimum value of the objective function is $W_N^* = \min W_N$ which is a quadratic function of X_{N-1}.

Next we consider the next period N-1. Note that by separability of the objective function and the state separation, the DP algorithm applies, hence we need only minimize the objective function

$$W_{N-1} = X_{N-1}^T M_{N-1} X_{N-1} - 2X_{N-1}^T M_{N-1} a_{N-1}$$
$$+ a_{N-1}^T M_N a_N + W_N^* \qquad (13.5)$$

with respect to U_{N-1} in period N-1. This objective function has two parts: the first is the contribution of stage N-1 and the second is the minimum cost W_N^* from the last period N. On using the value of W_N^* found in the previous stage, we get

$$W_{N-1} = X_{N-1}^T H_{N-1} X_{N-1} - 2X_{N-1}^T h_{N-1} + d_{N-1} \qquad (13.6)$$

where

$$H_{N-1} = M_{N-1} + A_N^T H_N (A_N + B_N G_N)$$

$$h_{N-1} = M_{N-1} a_{N-1} + A_N^T (h_N - H_N B_N g_N - H_N b_N)$$

$$d_{N-1} = a_{N-1}^T M_{N-1} a_{N-1} + (B_N g_N + b_N)^T H_N (B_N g_N + b_N)$$
$$- 2(B_N g_N + b_N)^T h_N + d_N .$$

Since the function (13.6) is of the same form as (13.3), we can repeat the process iteratively. Hence the optimal control policy at each period t is given by

$$U_t^* = G_t X_{t-1} + g_t \qquad t=1,2,\ldots,N \qquad (13.7)$$

where we can generate $(G_N, G_{N-1}, \ldots, G_1)$ and $(g_N, g_{N-1}, \ldots, g_1)$ sequentially as follows

Step 1. On using the initial condition $H_N = M_N$ we solve the set of equations (13.4) backward in time from t = N to t = 1 to yield the matrices $(G_N, H_{N-1}, G_{N-1}, H_{N-2}, \ldots, G_1)$.

Step 2. Next we use the initial condition $h_N = M_N a_N$ to rewrite the equations (13.6) backward in time from t = N to t = 1, thus yielding the vectors $(g_N, h_{N-1}, g_{N-1}, h_{N-2}, \ldots, g_1)$.

Step 3. Finally with G_t and g_t computed for t = N, N-1, \ldots, 2, 1 we compute the optimal control vector U_t^* by the linear decision rule (13.7).

In the general case of the LQG model, the linear dynamic system has a stochastic error term v_t

$$X_t = A_t X_{t-1} + B_t U_t + b_t + v_t$$

where the vector v_t is assumed to be normally distributed with zero mean for all t, fixed covariance matrix R and independence from $X_{t-1}, X_{t-2}, \ldots, U_t, U_{t-1}, \ldots$. The objective function is also changed to

$$\text{Min } E_o(W) = \text{Min } E_o[\sum_{t=1}^{N} (X_t - a_t)^T M_t (X_t - a_t)]$$

where E_o is the expected value conditional on the initial condition X_o. Again the optimal control rule takes the feedback form

$$U_t^* = G_t X_{t-1} + g_t, \qquad t=1,2,\ldots,N \tag{13.8}$$

where G_t, g_t are as before in (13.7). By certainty equivalence the optimal control law in the stochastic case is identical with that in the deterministic case. There are two subtle differences. One is that the optimal control U_t^* in the stochastic case can be calculated only after the state X_{t-1} is observed. If there are nonzero errors of observation or measurement in X_{t-1}, they would make the optimal control law imprecise and hence inoptimal. Second, the assumption of independence of the error v_t from past states $X_{t-1}, X_{t-2}, \ldots$ and past and current controls $U_t, U_{t-1}, U_{t-2}, \ldots$ is rarely fulfilled in empirical applications and the failure of this independence leads to optimal control rules more general than (13.8) and in particular they allow the estimate of the covariance term R to influence the choice of the optimal control rules. These methods are then called optimal adaptive control and they use techniques for updating the optimal control sequentially, which are known as Kalman-Bucy and other types of filtering techniques.

It is clear from the specification of the model (13.1), (13.2) that its format is identical with that of macro-dynamic Keynesian model, where the policymaker (e.g. the government) intends to minimize the sum of deviations from a desired level of national income say. Similarly the optimal production-cum-inventory models can be set up in this format.

PROBLEMS FOUR

1. In a two-region economy, national income y is the sum of two regional incomes y_1 and y_2 where $y_i = b_i x_i$, where x_i = capital stock of region i=1,2 with its constant output-capital ratio b_i. Let $u = u(t)$ be the proportion of total investment the planner allocates to region one and the remaining $(1-u)$ is for region two. The growth of capital in the two regions are

$$dx_1/dt = Dx_1 = u(g_1 x_1 + g_2 x_2)$$
$$dx_2/dt = Dx_2 = (1-u)(g_1 x_1 + g_2 x_2) \qquad (14.1)$$
$$g_i = s_i b_i \; (i=1,2), \; s_i = \text{constant regional savings ratio}.$$

The objective of the planner is to maximize

$$J = \int_0^T [\exp(-\delta t) \, c(t)] dt$$

a discounted stream of consumption $c = (1-s_1)b_1 x_1 + (1-s_2)b_2 x_2$ over a fixed horizon T, subject to the above conditions (14.1) on growth of regional capital and the restriction that $u(t)$ is continuous with $0 \leq u(t) \leq 1$ for all t in $0 \leq t \leq T$. Derive the necessary conditions for the optimal solution paths by Pontryagin's maximum principle and show that the optimal control follows a bang-bang control.

2. Solve the intertemporal minimization problem

$$\underset{u}{\text{Min}} \int_0^T dt$$

s.t. $dx_1/dt = x_2$, $dx_2/dt = u$
$|u(t)| \leq 1$

where there are two state variables x_1, x_2 and one control variable u. Show that the optimal control has the bang-bang form

$$u^*(t) = \begin{cases} 1 & \text{if } c_2 - c_1 t \geq 0 \\ -1 & \text{if } c_2 - c_1 t < 0 \end{cases}$$

where c_1, c_2 are certain constants.

CHAPTER FOUR

3. The Arrow-Karlin production model minimizes the sum of production and inventory costs over a fixed planning horizon T

$$\min_{u(s) \geq 0} J = \int_0^T [C(u(s)) + H(x(s))]ds$$

s.t. $dx/dt = u(t) - r(t)$

$x(0) = x_0$ given initial inventory

where $x(t) = x_0 + \int_0^t [u(s) - r(s)]ds \geq 0$ is the cumulative inventory, $u(t)$ is current production and $r(t)$ is current demand known or estimated. Assume that $\partial C(u)/\partial u > 0$ and $\partial H(x)/\partial x > 0$ and C and H are strictly convex, then derive the necessary conditions to be satisfied by the optimal solution paths.

4. Given the initial state $x(0) = x_0$, the terminal time T and the dynamics of the single state variable $x(t)$ as

$$dx/dt = a\, x(t) + b\, u(t)$$

where a, b are fixed constants, we want x to return to the origin $x = 0$, where the origin is the desired state. The cost is:

$$C = \int_0^T u^2(t)dt + x^2(T)$$

where the first term is the cost of control, the second being the cost of deviation from the origin. Show that the optimal control $u^*(t)$ varies exponentially with time as

$$u^*(t) = -x(T)\, b\, \exp(a(T - t))$$

5. A given positive quantity x is to be divided into n parts such that the product of the n parts is a maximum. Formulate the model as a maximization problem and solve by the dynamic programming algorithm. Show that the optimal solution is to choose the n parts such that each part equals x/n.

6. The steady state solution of the infinite horizon neoclassical growth model

$$\text{Max } J = \int_0^\infty U(c) \exp(-\delta t)dt$$

s.t. $dk/dt = f(k) - (n+r)k - c$

k_0 given

has been criticized on two grounds: it may take very long periods to reach the steady state and the discount rate δ is arbitrary. Discuss these criticisms as they affect the stability of the optimal path of capital accumulation.

7. For the deterministic LQG model

$$\text{Min } J = \int_s^T [u^2 + a(x_T - x)^2] dt$$

s.t. $dx/dt = u$

x_0, x_T fixed and $a > 0$

show that the optimal profile of the adjoint variable $p = p(t)$ i.e. dynamic shadow price satisfies the following time path

(i) $p(t) = A_1 \exp(\lambda_1 t) + A_2 \exp(-\lambda_1 t)$ for finite T
(ii) $dp(t)/dt = -A_2 \lambda_1 \exp(-\lambda_1 t)$ for infinite T

where A_1, A_1 are two constants to be determined and λ_1, λ_2 are the two roots of $\lambda^2 = a$ i.e. $\lambda_1 = \sqrt{a}$, $\lambda_2 = -\sqrt{a} = -\lambda_1$.

8. For the three period maximization problem

$$\max_{u_i \geq 0} J = \sum_{i=1}^{3} u_i^2$$

s.t. $x_{i-1} = x_i - u_i$ $i = 1,2,3$

$x_i = 0,1,2,\ldots,5$ (integers)

$x_3 = 5$

show that the optimal solution satisfies the following functional equations by the DP algorithm

$f_1(x_1) = \max u_1^2$, $u_i = 0,1,\ldots,x_i$
$f_2(x_2) = \max [u_2^2 + (x_2 - u_2)^2]$, $u_i = 0,1,\ldots,x_i$
$f_3(x_3) = \max [u_3^2 + f_2(x_3 - u_3)^2]$, $u_3 = 0,1,2,3,4,5$.

9. For the neoclassical model with a fixed horizon T, assume that the production function is $f(k) = k^a$, $a > 0$. Derive the second order nonlinear differential equation for $k(t)$ satisfied by the optimal trajectory. Linearize this equation around the steady state value k^* and solve it.

10. Discuss the statement that a dynamic intertemporal path contains for each t a short run path.

CHAPTER FIVE
STATIC AND DYNAMIC GAMES

Any discussion of economic models is incomplete, if it does not contain an analysis of game theory and its various applications. This is because the models in game theory provide a substantial amount of generalization of most of the basic mathematical concepts we have so far used e.g., the concept of equilibrium, stability, saddle point and intertemporal optimum. A game may be viewed as a team decision problem, when the different members of the team have cooperative or non-cooperative (i.e. conflicting) objectives to optimize, with different or equal information structures and finite or infinite strategies to choose from. Four general ways to classify games are: (1) static or dynamic, where the latter involves intertemporal optimization by different players with strategies selected sequentially over time, (2) deterministic or stochastic, where for the latter it is assumed that there exist stochastic errors in the objective function or the constraints on the strategy choice, which may be due to incomplete information available to each player, (3) games of strategy (i.e. active game against rational opponents), or games against nature (i.e. passive game, where the opponent is viewed as nature, who is not assumed to have any motives of gain or retaliation), and (4) the three analytical forms of the game model: the extensive form, the normal form and the characteristic function form. The normal form is most frequently applied in two-person games, whereas the characteristic function form is most useful for n-person games with n greater than two.

5.1 Forms of Games
5.1.1 Extensive Form

An n-person game in extensive form may be viewed as a finite tree, called <u>game tree</u> with a finite collection of nodes, called <u>vertices</u>, connected by lines called <u>arcs</u>, so as to form a connected figure. It has the following characteristics:

(a) a particular node of the game tree T referred to as the starting point, also called the first move,

(b) a function, called the <u>payoff function</u> defined for each terminal vertex of the tree, which may be viewed as an n-tuple vector of rewards, with one element for every player,

(c) a partition of the nonterminal vertices of the tree T into n+1 sets $S_0, S_1, S_2, \ldots, S_n$ called the player partitions, the nodes in S_0 are called <u>chance moves</u> (i.e., moves selected from some probability distributions) and the nodes in S_i

STATIC AND DYNAMIC GAMES

are called moves of player i, for i=1,2,...,n and

(d) for each i ∈ N, where N = {1,2,...,n} is the set of players, a sub-partition of each S_i into subsets $S_i(j)$, j ∈ {1,2,...,k_i} called the <u>information set</u> of player i, such that two nodes in the same information set have the same number of outgoing branches and there is a one-to-one correspondence between the sets of outgoing branches of different nodes in $S_i(j)$ and no node can follow another node in the same information set.

Thus there are four basic elements of a game in extensive form: a starting point, a payoff function, a set of chance and personal moves for each player and an information set for each player, who knows what information set he is in but not which vertex or node of the information. A simple example of an extensive form game is that of matching pennies as follows:

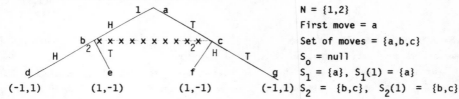

N = {1,2}
First move = a
Set of moves = {a,b,c}
S_o = null
S_1 = {a}, $S_1(1)$ = {a}
S_2 = {b,c}, $S_2(1)$ = {b,c}

The starting point is at the vertex marked a, where the first move by player 1 begins. Player 1 chooses heads H or, tails T. Player 2 selects H or T not knowing player 1's choice. If the two choices are alike HH or, TT then player 2 wins a penny from player 1; otherwise player 1 wins a penny from the second player. In the game tree shown, the 2-tuple vectors at the four terminal nodes d, e, f, g denote the four payoff functions, the numbers near the other nodes refer to the player to whom the moves correspond and the crossed area bc incorporates moves belonging to the same information set.

Three concepts are most important in this characterization. One is the notion of <u>perfect</u> or <u>imperfect information</u>. A player i is said to have perfect information in a game tree T if his information sets $S_i(j)$ consist only of one element. If each player has perfect information, then the whole game T is said to have perfect information, otherwise we have imperfect information. For example chess is a perfect information game but poker is not. The second notion is that of a <u>strategy</u> for each player i. This is the choice plan for playing a game; mathematically it specifies the choice of one of the branches outgoing from a specific node contained in the information set $S_i(j)$ of player i. The set of all strategies of player i in the entire game T may be denoted by T_i. The third important concept underlying an extensive form game is that each player is aware of the rules of the game and the payoff functions of each of the players but since there are chance moves, no one knows except probabilistically what the results of the chance moves will be on the actual realized payoffs (or outcomes). Hence it is assuemd that each player chooses his strategy so as to maximize the expected payoff function or the expected utility of the outcomes. Thus from the given payoffs of the game in extensive form, we may define a payoff function defined over n different variables, the i-th one representing the strategies available to player i=1,2,...,n. First, assume there are no chance moves in the extensive game, so that a particular selection of strategies $(s_1, s_2, \ldots, s_n)$, one chosen by each player determines a play denoted by α. Then we define player i's payoff or utility for its strategy-mix as $U_i(s_1, s_2, \ldots, s_n) = U_i(\alpha)$. When there are chance moves, then each selection of the strategy-mix $(s_1, s_2, \ldots, s_n)$ occurs with a probability. Denote by $p(\alpha)$ the probability of play α occurring. Then for the payoff associated with the strategies $s = (s_1, s_2, \ldots, s_n)$ we take the expected value over all the plays i.e.

$$U_i(s_1, s_2, \ldots, s_n) = \sum_{\alpha=1}^{K} p(\alpha) U_i(\alpha) \tag{1.1}$$

where it is assumed that there are K plays. Payoffs to each player measured this way are said to satisfy the conditions of a von Neumann-Morgenstern utility function U_i for each player i. Note that this notion helps to simplify considerably the extensive form game even with chance moves. This simplified game, which is called a <u>normal form</u> has the following structure: each player i now has exactly one move out of his choice set with his expected payoff function $U_i(s)$ and he makes his choice of s_i without exactly knowing the strategy choices by other players. In the game of matching pennies, each player has two strategies H and T. The normal form of this game is illustrated by the 2x2 payoff matrix

		player 2	
		H	T
player 1	H	(-1,1)	(1,-1)
	T	(1,-1)	(-1,1)

where each row denotes a strategy of player 1 and each column a strategy of player 2. Thus for a game T in normal form with n players we have three elements: (i) n players (ii) the pure strategy sets (i.e. when there are no chance moves) are $T_1, T_2, \ldots, T_n$ and (iii) the payoff functions are $U_1, U_2, \ldots, U_n$ which have to be interpreted as expected payoffs or expected utilities whenever necessary.

The notion of each player maximizing his expected utility i.e. by choosing that strategy which yields the highest expected payoff has several powerful implications. First, it allows a method of reducing the uncertainty of the situation into a non-stochastic form by using either the subjective probability assumption of each player or the objective knowledge. Second, it leads to a normative behavior which can sometimes be empirically tested against observed behavior. Third, it yields one of the most important concepts in game theory i.e. the <u>equilibrium</u> point or strategy-mix. An n-tuple vector $s^* = (s_1^*, s_2^*, \ldots, s_n^*)$ of strategies is said to be in equilibrium in a game T, if and only if for any i and any other $s_i \in T_i$ we have

$$U_i(s_1^*, s_2^*, \ldots, s_{i-1}^*, s_i, s_{i+1}^*, \ldots, s_n^*) \leq U_i(s_1^*, \ldots, s_n^*)$$

In other words the equilibrium strategy vector $s^* = (s_1^*, s_2^*, \ldots, s_n^*)$ is preferred over any nonequilbrium ones, since no player has any incentive to choose lower payoffs. Thus an equilibrium point has stability in the sense that once found, the players may tend to stick to it, provided each player knows what the others will play. As an example consider the two player game in normal form

	y_1	y_2	
x_1	(5,4)	(3,3)	player 1 strategies: (x_1, x_2)
x_2	(3,3)	(4,5)	player 2 strategies: (y_1, y_2)

Here both (x_1, y_1) and (x_2, y_2) are equilibrium points.

5.1.2 Normal Form

A game in normal form which is most commonly used is a two-person game, where the payoffs of the two players may add up to zero (zero-sum game) or, a constant (constant sum game). The former is also called a noncooperative or strictly competitive game, since by $U_1 = -U_2$ the objective of one player is the direct opposite of that of his rival. In two-person zero-sum games it is most useful to characterize an equilibrium pair of strategies, if it exists. This is so because it may imply in some sense a degree of stability.

Suppose we do not have chance moves and each player chooses one of a finite set of his strategies. Let the payoff matrix for the zero-sum game be

	y_1	y_2	y_3	Min row	
x_1	8	4	6	4	
x_2	6	5*	7	5	saddle point: (x_2, y_2)
x_3	0	3	4	0	payoff (5*,-5*)
Max column	8	5	7		

where the rows x_i (columns y_j) denote the strategies of player one (two) and the elements (a_{ij}) of the payoff matrix denote player one's gain, the other player's gain being $(-a_{ij})$ by the zero-sum property e.g., for the pair (x_1, y_3) of strategies player 1 gains 6 from player 2 but player 2 loses 6 (i.e., gains -6) to player 1. Clearly, the pair (x_2, y_2) resulting in payoffs (5,-5) is an equilibrium pair, since player 1 may get less than 5 if he chooses x_1 (with y_2 chosen by his rival) or, x_3 (with his rival choosing y_1); likewise player 2 may lose more than 5 if he chooses

y_1 (with x_1 selected by his rival), or y_3 (with x_2 set by his opponent). This type of equilibrium point is called a <u>saddle point</u> by analogy with the surface of a saddle which curves upwards in one direction (e.g., player 1 maximizes his gain) and downward in the other direction (e.g., player 2 minimizes his loss). The pair (x_2, y_2) here is called a saddle point in <u>pure strategies</u>, since only one strategy has to be chosen by each player.

But if the game is played many times say 100 times, then player 1 may choose the three strategies 30%, 50% and 20% of the times. This mixing of pure strategies allows us to define <u>mixed strategies</u>, which have two interesting interpretations. One is that in repeated play of the game, this allows each player to mix the set of pure strategies which are discrete i.e., x_1, x_2, x_3 become the proportion of times player 1 chooses the three strategies $x_i \geq 0$, $x_1 + x_2 + x_3 = 1$ and likewise for player 2, $y_j \geq 0$, $y_1 + y_2 + y_3 = 1$. A second reason is that the saddle point equilibrium may not exist for many finite games, if only pure strategies are allowed. For example the matrix game with the payoff

	y_1	y_2	Min row
x_1	8	6	6
x_2	5	7	5
Max column	8	7	

has obviously no saddle point in pure strategies. It has however a saddle point in mixed strategies. To find this mixed strategy saddle point we interpret x_i, y_j as probabilities i.e., percent of the times strategy i is chosen by player 1 and strategy j by player 2. If R_1 and R_2 denote the expected payoff of player 1 when he chooses strategy 1 and 2 respectively, then

$$R_1 = 8y_1 + 6y_2 = 8y_1 + 6(1 - y_1), \text{ since } y_2 + y_1 = 1$$
$$R_2 = 5y_1 + 7y_2 = 5y_1 + 7(1 - y_1), \text{ since } y_2 = 1 - y_1.$$

If he is to be indifferent between R_1 and R_2 (because he can also play a mixed strategy), then we must have $R_1 = R_2$ for player 1. This yields $y_1 = y_1^* = 1/4$, $y_2 = 1 - y_1 = y_2^* = 3/4$. Similarly, if L_1 and L_2 are the expected loss of player 2, when he selects strategies 1 and 2 respectively, then

$$L_1 = 8x_1 + 5x_2 = 8x_1 + 5(1 - x_1), \text{ since } x_1 + x_1 = 1$$
$$L_2 = 6x_1 + 7x_2 = 6x_1 + 7(1 - x_1), \text{ since } x_2 = 1 - x_1.$$

By equating $L_1 = L_2$ we obtain $x_1 = x_1^* = 1/2$, $x_2 = x_2^* = 1/2$. The expected gain of player one is 6½, which is also the expected loss of player 2 when the above mixed strategies (x^*, y^*) are chosen where $x^* = (x_1^*, x_2^*)$, $y^* = (y_1^*, y_2^*)$. There is a third way of finding out the common payoff 6½ denoted by z^* say where

$$z^* = \sum_{i=1}^{2} \sum_{j=1}^{2} x_i^* a_{ij} y_j^* \qquad (1.2)$$

with $A = (a_{ij})$ being the two-by-two payoff matrix above i.e.

$$z^* = 8x_1^* y_1^* + 6x_1^* y_2^* + 5x_2^* y_1^* + 7x_2^* y_2^*$$

$$= 8 \cdot \frac{1}{2} \cdot \frac{1}{4} + 6 \cdot \frac{1}{2} \cdot \frac{3}{4} + 5 \cdot \frac{1}{2} \cdot \frac{1}{4} + 7 \cdot \frac{1}{2} \cdot \frac{3}{4} = 6\frac{1}{2}.$$

Several features of this mixed-strategy saddle point solution must be noted. First of all, each player is doing better than the worst possible situation e.g., player 1's worst outcome is 5 for strategy x_2 and player 2's worst outcome is the loss of 8 for y_1 but the saddle point solution is $z^* = 6\frac{1}{2}$, so that player 1 gains on the average $1\frac{1}{2}$, while player 2's loss is reduced on the average by $1\frac{1}{2}$. Second, the mixed strategy saddle point $(x^*, y^*; z^*)$ provides a safety first rule for each player and a pure strategy saddle point is included in a mixed strategy, provided the former existed. A pure strategy saddle point is often called 'the best of the worst' strategy or solution. As we saw in Chapter Two every optimal (feasible) solution of a finite linear programming (LP) model can be viewed as a mixed strategy saddle point in a two-person zero-sum game and vice versa. Thirdly, a mixed strategy saddle point can be considered as a starting point for many bargaining situations, where the final solution may be different. Lastly, the saddle point in two-person zero-sum games is very closely related to the Kuhn-Tucker theory applied to linear and nonlinear programming.

5.2 Two-person Zero-sum Games

The following three types have been most widely applied in economic theory of two-person games: (1) game against nature as a passive opponent, (2) game of strategy against an active opponent who may retaliate and (3) the differential game, when the objectives are intertemporal and dynamic strategies over time are to be selected.

We illustrate these three types of two-person games (TPG), which may have zero, constant or increasing sum of payoffs, although the zero-sum case would be used most.

5.2.1 Game Against Nature

This is a passive type of TPG, which is most suitable for the decision problem known as decision-makng under uncertainty, when the decisions are influenced by the alternative states of nature. A typical example is the farmer's decision to fertilize (action 1), or not to fertilize (action 2), when for example there are only two states of nature: rain (state 1) and no rain (state 2). Assume that the net income of the farmer in dollars is given by the payoff matrix $A = (a_{ij})$

Actions	States of nature	
	1	2
1	60	20
2	40	30
Probability	0.5	0.5

$A = (a_{ij})$:   (2.1)

Here we intepret each payoff conditional on a given action and a given state of nature e.g., the farmer obtains $20 if he fertilizes but there is no rain. Which action the farmer should adopt? The answer depends on the amount of information he has about the states of nature and the degree of risk he is willing to take. If there is not much past data on which he can reliably estimate the probability of rain by following the acceptable statistical criteria, we have a case of complete uncertainty. In this case the information provided by the last row marked by probability giving the probabilities of rain and no rain is not available. This is the case of complete ignorance or uncertainty. A slightly better situation is when some data are available but not as much as is needed for a good estimate. How should the farmer revise his actions in the light of new data? It is clear that such decision problems do frequently arise when a company has to choose between two or more types of advertising the product it sells and the states of nature represent strong or weak market response. In this case the company may even incur some cost to get sample information through sample surveys or opinion polls to assess the relative probabilities of the states of nature.

Three cases are usually distinguished about the states of nature: (a) a state of certainty when there is perfect information as to which state is going to occur, (b) a state of risk when perfect information is not available but the probabilities about the different states can be estimated from past data and (c) a state of uncertainty where the probabilities of the different states of nature are unknown. In the third case various decision criteria have been proposed, which may be illustrated with respect to the farmer's payoff matrix (2.1) mentioned before

(1) Maximin criterion, also called the minimax criterion if the payoff is viewed as a loss function which prescribes a strategy yielding the best of the worst outcomes. Thus for each action we compute the minimum or worst payoff and then choose the best or the maximum

Actions	Worst payoff (min)	Best (max)
1	20	
2	30	30

clearly action 2 is recommended by this criterion, which is also called the Wald criterion after its originator Abraham Wald.

(2) Bayes or Laplace criterion, also called the principle of insufficient reason, which assigns equal probabilities to the different states of nature and then recommends the strategy whose expected payoff is the highest. Thus for two states

of nature, we have 0.5 as the probability of each state of nature and the expected payoffs are 0.5(80) = 40 and 0.5(70) = 35 for actions 1 and 2 respectively. Hence action 1 is recommended.

This criterion has two important features. One is that the probabilities may be interpreted subjectively i.e. the procedure requires the decision-maker to assign subjective probabilities to nature's possible strategies. These subjective probabilities are then the <u>prior probabilities</u> of states of nature in the language of statistical decision theory. With more information or more sampling experiments, an updated or revised version of these probabilities may be available later on and used by the decision-maker. This updated probability distribution is called the posterior distribution and depending on the form of the distribution and new sampling observations, these <u>posterior probabilities</u> can be precisely calculated. Secondly, this criterion is most closely related to the von Neumann-Morgenstern theory of maximizing the expected utility of the payoff. Thus, if $u_i(a_{ij})$ is the utility of the payoff a_{ij} for fixed i and varying states of nature j, then the expected utility for action or strategy i (i=1,2,...,m) is $\sum_{j=1}^{n} u_i(a_{ij}) y_j = U_i$, if y_j is the prior probability of state j=1,2,...,n. The decision maker chooses that strategy i=k for which U_i is the maximum over all i=1,2,...,m i.e. $\max_{1 \leq i \leq m} U_i$ yields the optimal strategy x_k say.

(3) Hurwicz (optimism-pessimism) criterion, due to Leonid Hurwicz defines for each action i a weighted average of the maximum (i.e. best) and the minimum (i.e. worst) payoff i.e. $R_i = w \max_{1 \leq j \leq n} a_{ij} + (1-w) \min_{1 \leq j \leq n} a_{ij}$ with $0 \leq w \leq 1$ and then chooses that strategy for which R_i is highest for a given value of w, which is termed the coefficient of optimism. Note that the closer w is to 1, the more optimistic the decision maker is. For our example (2.1) with w = 0.4, R_1 = 36, R_2 = 34, hence action 1 is optimal by this criterion, provided the decision maker has the optimism index given by w = 0.4. Note that for a low optimism index such as w = 0.2, the optimal action is action 2.

(4) Minimax regret criterion due to L.J. Savage defines for each action i the amount of regret or opportunity loss $L_i = \max_{1 \leq j \leq n} a_{ij} - a_{ij}$ by subtracting alternative payoffs for particular states of nature from the maximum payoff and then requires that the minimum of the maximum regrets be selected. For our example (2.1) the regret matrix is:

```
0    40
0    10
```

hence action 2 with minimax regret of 10 is recommended.

(5) Mixed strategy criterion computes the expected payoff $R_j = \sum_{i=1}^{m} x_i a_{ij}$ for each pure strategy j of nature by using a random device (e.g. a coin or dice) to define the proportions x_i ($x_i \geq 0$, $\sum_{i=1}^{m} x_i = 1$) and then selects the x_i's so that the minimum value $v = \min_{1 \leq j \leq n} R_j$ is maximized. If v is taken as the safety level, then this minimax mixed strategy criterion maximizes the safety level of payoff. Clearly this is identical with the selection of the odds (i.e. optimal x's) which yield the lowest minimax regret. The optimal mixed strategy solution $x^* = (x_1^*, x_2^*)$ may thus be obtained by solving the following LP problem

Max v   subject to
 x

$$\sum_{i=1}^{m} x_i a_{ij} \geq v, \quad \sum_{i=1}^{m} x_i = 1, \quad x_i \geq 0 \qquad (2.2)$$

$$i=1,2,\ldots,m$$
$$j=1,2,\ldots,n.$$

On using y_j and u as the Lagrange multipliers, the dual LP model can be formulated as

Min u   s.t. $\sum_{j=1}^{n} a_{ij} y_j \leq u, \quad \sum_{j=1}^{n} y_j = 1, \quad y_j \geq 0 \qquad (2.3)$
 y

$$j=1,2,\ldots,n.$$

Since the sets $\{x \mid \sum_i x_i = 1, x_i \geq 0, i=1,2,\ldots,m\}$ and $\{y \mid \sum_{j=1}^{n} y_j = 1, y_j \geq 0,$ $j=1,2,\ldots,n\}$ are convex closed bounded and nonempty, there must be feasible vectors x and y and hence optimal feasible vectors if the payoff matrix has finite number of strategies for each player. Let (x^*, v^*) and (y^*, u^*) be the optimal pair of solutions of the two respective LP problems (2.2) and (2.3), then by the duality theorem $u^* = v^*$ i.e. minimax loss of player 2 (nature) = maximin gain of player 1 (decision-maker). For our example (2.1), the optimal solutions are

$x_1^* = 0$, $x_2^* = 1$, $v^* = 30$ and $y_1^* = 0$, $y_2^* = 1$, $u^* = 30$.

Suppose the payoff matrix were different e.g.,

50	30
20	40

and suppose nature uses a completely mixed optimal strategy so that y_j^* is greater than zero but less than one for j=1,2, then by the complementary slackness property of LP we get

$$R_1^* = 50x_1^* + 20(1 - x_1^*) = R_2^* = 30x_1^* + 40(1 - x_1^*)$$

yielding $x_1^* = 1/2 = x_2^*$, $R_1^* = v^* = 35$.

Similarly we equate nature's expected losses $L_1^* = L_2^*$ to get

$$L_1^* = 50y_1^* + 30(1 - y_1^*) = L_2^* = 20y_1^* + 40(1 - y_1^*)$$

giving $y_1^* = 1/4$, $y_2^* = 3/4$, $L_1^* = u^* = 35$.

As mentioned before the mixed strategy criterion has a degree of pessimism built into it, since it postulates first the worst situation the decision maker may be put to by nature and then he chooses a mixing of pure strategies so as to have the best of the worst in solving for the LP model (2.2). Also, the mixed strategies x_i^* (i=1,2,...,m) are probabilities which can be realized only in several rounds of the play; it is not clear how many rounds of the play are needed in a given case; in statistical language this is equivalent to specifying how large a sample should be for the large sample theory to apply.

Some comments are now in order for the several decision criteria presented above. Clearly, with so many criteria suggesting different actions to be optimal, the decision maker has the dilemma of choosing one criterion and for further analysis must appeal to basic principles of rationality. However, the existing philosophical theories of reason does not provide any definitive guidelines. However, in statistical applications of applied decision theory two criteria have often been used: the Bayesian criterion and the mixed strategy saddle point. The first has been used because prior probabilities may be revised or updated with more information and experience. Bayesian procedures can be systematically applied in these updating schemes and recently empirical Bayes methods are often applied whereby the updating process is considerably improved.

The main appeal of the mixed strategy solution is that it allows the choice of a cautious safety-first policy and it is in keeping with the expected utility theorem of von-Neumann and Morgenstern in the following sense

$$\text{Max}_x v, \quad \text{s.t.} \quad \sum_{i=1}^{m} x_i u_i(a_{ij}) \geq v$$

$$\Sigma x_i = 1, \quad x_i \geq 0, \quad i=1,2,\ldots,m$$

where $u_i(a_{ij})$ is the utility of the conditional payoff a_{ij} given action i. The expected utility theorem says that for a decision maker who accepts certain reasonable postulates of coherent preference there exists a utility function, the

192   CHAPTER FIVE

expected value of which can be used by the individual to choose among risky (uncertain) prospects. Thus the payoffs a_{ij} may be stochastic rather than deterministic, still one could define a mixed strategy solution.

5.2.2 Cournot Market Games

As a two-person game of strategy, the duopoly market model due to Augustin Cournot, which has been generalized by Nash to cover bargaining situations, is most widely applied in various situations both static and dynamic. This is because the Cournot-Nash games can be generalized in several useful ways e.g., (1) cooperative and noncooperative games, (2) possibility of threats on both sides, (3) extension to n-player games and (4) stochastic payoffs due to market response not completely known.

Cournot's simple model developed in the 1830s considered a homogeneous product, water to be supplied by two producers from a mineral spring each with a zero cost of production. The total market demand for mineral water is $p = 1 - (x_1 + x_2)$, where x_i is the output of producer i=1,2. Clearly, the profit of each producer i is $z_i = x_i(1 - x_1 - x_2)$ for i=1,2. The question is: What output will the two producers produce? Cournot postulated a solution (x_1^*, x_2^*), which is called Cournot-equilibrium or, Cournot-Nash (CN) equilibrium which satisfies the following three postulates: (1) each x_i^* maximizes the payoff or profit z_i of player or, producer i, given a fixed value of the other player's strategy or output x_j^*, (2) total output $x_1^* + x_2^*$ clears the market resulting in a meaningful price i.e. a positive equilibrium price $p = 1 - (x_1^* + x_2^*)$, and (3) each firm changes its output x_i till the first two conditions are met and there is no reaction from his rival. The first is called the rationality postulate, the second a market clearing condition of equilibrium and the third a reciprocity condition of passive reaction, also known as zero conjectural variation or zero degree of retaliation. Note that the third postulate merely prescribes a method of sequential revision of outputs x_i till the equilibrium outputs x_1^*, x_2^* are reached for each player. Two major objections to this postulate which is most important in Cournot theory are that it is very unrealistic and perhaps unnecessary. As a matter of fact this had led to the development of a host of other reasonable solutions different from Cournot's. It is necessary to point out however, in all fairness to Cournot that the third postulate on the two reaction curves of the players only describes a method of reaching the equilibrium solution (x_1^*, x_2^*); if the latter exists, one may certainly devise other methods of realizing it or converging to it. In other words, the reaction curve technique is merely a computational algorithm.

We now consider the various alternative solutions in relation to the mineral water example above.

(1) Cournot solution

Here player i maximizes his own profit

$$z_i = x_i(1 - (x_1 + x_2)), \quad i=1,2$$

assuming the other player's strategy to be given or estimated. On setting $\partial z_i/\partial x_i$ to zero we obtain the two optimal reaction curves

$$x_1 = (1 - \hat{x}_2)/2, \quad x_2 = (1 - \hat{x}_1)/2 \tag{3.1}$$

where $\hat{x}_2$ and $\hat{x}_1$ are the guesses made by player 1 and 2 respectively about the rival's strategy. It is easy to check that the above first order conditions (3.1) are also sufficient for the maximum. Note that these linear reaction curves (3.1) $x_1 = x_1(\hat{x}_2)$ and $x_2 = x_2(\hat{x}_1)$ are only optimal for the given guessed values $\hat{x}_2, \hat{x}_1$. But the guesses are not at equilibrium till we have $x_1 = \hat{x}_2 = x_1^*$, $x_2 = \hat{x}_2 = x_2^*$ simultaneously. Hence we set these conditions and solve for (x_1^*, x_2^*) from (3.1) as $x_1^* = 1/3 = x_2^*$. It is easy to show that the strategy sets for the two players are: $X_1 = \{x_1: 0 \leq x_1 \leq 1\}$, $X_2 = \{x_2: 0 \leq x_2 \leq 1\}$ and the CN equilibrium (x_1^*, x_2^*) satisfies the optimality condition for the profits z_i

$$\begin{aligned} z_1(x_1^*, x_2^*) &> z_1(x_1, x_2^*), \text{ all } x_1 \in X_1 \\ z_2(x_1^*, x_2^*) &> z_2(x_1^*, x_2), \text{ all } x_2 \in X_2 \end{aligned} \tag{3.2}$$

where $z_1(x_1^*, x_2^*) = 1/2 = z_2(x_1^*, x_2^*)$.

Also for any guesses $\hat{x}_2$ in X_2, the optimal profit of player 1 is $z_1(x_1, \hat{x}_2) = (1 - \hat{x}_2)^2/4$; likewise $z_2(\hat{x}_1, x_2) = (1 - \hat{x}_1)^2/4$ for the guess of $\hat{x}_1$ in X_1 made by player 2. Hence the worst and the best situation for player 1 is 0 and 1/4 for $\hat{x}_2 = $ 1 and 0. Similarly for player 2. Clearly the CN equilibrium solution $(x_1^*, x_2^*) = $ (1/9, 1/9) is intermediate between the worst and the best profit situation for each player.

Suppose x_2 is fixed at x_2^* and so announced by player 2 and then player 1 maximizes his own profit $z_1(x_1, x_2^*) = x_1 - x_1^2 - x_1 x_2^*$ by selecting his output level x_1 in his strategy set. This yields $x_1 = (1 - x_2^*)/2 = x_1^*$. Likewise if player 1 preannounces his strategy at $x_1 = x_1^*$, player 2 would attain his maximum profits at $x_2 = (1 - x_1^*)/2 = x_2^*$. Since the second derivatives of $z_1(x_1, x_2^*)$ for given x_2^* and of $z_2(x_1^*, x_2)$ for given x_1^* are negative, the convergence of x_i to x_i^* is assured by a trial and error process. This may also be shown by checking that the two linear reaction curves in (3.1) written as

$$2x_1 + x_2 = 1, \quad x_1 + 2x_2 = 1$$

intersect at the point $x_1 = x_2 = x_i^* = 1/3$, $i=1,2$.

Two interesting implications of the Cournot model are that it is easily dynamized and easily applicable to an n-player situation with the same set of assumptions. To illustrate the first case, assume a dynamic demand function

$$p = 1 - (x_1 + x_2) - (Dx_1 + Dx_2), \quad Dx_i = dx_i/dt$$

and that each player is maximizing an intertemporal profit function with fixed initial and terminal points $x_i(0) = A_i$, $x_i(T) = B_i$

$$\text{Max } J_i = \int_0^T z_i \, dt$$

where

$$z_i = px_i, \quad i=1,2.$$

By applying the Euler-Lagrange equations for a maximum for each $i=1,2$ we obtain a pair of linear differential equations to be satisfied by the optimal trajectory

$$dx_1/dt + x_1 + 2x_2 = 1$$
$$dx_2/dt + 2x_1 + x_2 = 1.$$

We solve these differential equations subject to the two boundary conditions for each x_i. It is easy to check from the characteristic equation $\lambda^2 + 2\lambda - 3 = 0$ that the two roots are $\lambda_1 = 1$, $\lambda_2 = -3$ real and opposite in sign. Hence it defines a saddle point equilibrium. Also, the particular solution is given by the vector x^*

$$x^* = \begin{pmatrix} x_1^* \\ x_2^* \end{pmatrix} = \begin{bmatrix} 1 & 2 \\ 2 & 1 \end{bmatrix}^{-1} \begin{pmatrix} 1 \\ 1 \end{pmatrix} = \begin{pmatrix} 1/3 \\ 1/3 \end{pmatrix}$$

which is identical with the static Cournot solution.

In the n player case assume the demand and total cost functions as

$$p = a_1 - bS, \quad S = \sum_{j=1}^{n} x_j$$

$$C(x_i) = c_i + c\, x_i, \quad i=1,2,\ldots,n$$

where each player is assumed to have identical marginal costs (this assumption is needed to prevent the dominance of any one player in terms of cost advantage). Maximizing profits $z_i = px_i - C(x_i)$ for each player's output x_i by assuming others' outputs to be given constants we obtain the optimal reaction lines

$$x_i = (a/b) - S, \quad \text{for each } i=1,2,\ldots,n$$
$$a = a_1 - c > 0$$

Since this must hold for each $i=1,2,\ldots,n$ we can sum both sides over $i=1$ to $i=n$ and obtain

$$\sum_{i=1}^{n} x_i = S = n(a/b) - n S.$$

This yields the equilibrium values x_i^*, $S^* = \Sigma x_i^*$

$$S = S^* = \frac{na}{b(n+1)}, \quad x_i = x_i^* = \frac{a}{b(n+1)}, \quad i=1,2,\ldots,n.$$

As $n \to \infty$, the Cournot equilibrium solution tends to the competitive (or quasi-competitive) one where $\lim_{n \to \infty} S^* = \frac{a}{b} > \frac{na}{b(n+1)}$ for all $n > 0$ and the price equals marginal cost (p=c). This implication is very important in general equilibrium theory, since it shows that a duopolistic or oligopolistic market where the players have noncooperative or even conflicting objectives tend to converge to the competitive soltuion with higher aggregate output and lower price (i.e. it increases the social welfare) as the size of the market increases through n. Much of the argument behind anti-trust laws is based on this type of rationale.

(2) <u>Stackelberg solution</u>

This solution assumes that one of the two players is the leader and the other the follower. A follower say player 2 obeys his reaction curve i.e.

$$x_2 = (1 - x_1)/2$$

by adjusting his output in the Cournot way. A leader say player 1 does not obey his reaction curve as in Cournot model but assumes that his rival acts as a follower and substitutes his rival's reaction function above into his own profit function i.e.

$$\begin{aligned} z_1(x_1, x_2) &= x_1(1 - x_1 - x_2) \\ &= x_1(1 - x_1 - (1-x_1)/2) \end{aligned}$$

and then maximizes it with respect to his own output level x_1. This yields from $\partial z_1(x_1,x_2)/\partial x_1 = 0$ the Stackelberg solution $x_1 = 1/2$ and hence $x_2 = (1 - x_1)/2 = 1/4$ with profits $z_1 = 1/8$, $z_2 = 1/16$. There are two sources of instability in this approach. One is that each may like to be the leader rather than the follower and as long as there is no specific cost advantage for any one, this may only be determined by negotiation or bargaining; otherwise the battle may continue, since the follower may not accept a profit level lower than the leader's. Secondly, the reaction function of the follower may only be imperfectly known or even unknown to the leader, in which case the solution is indeterminate. Also, it is not very easy to generalize to the case of n players.

(3) **Pareto solution**

This is a cooperative solution obtained by choosing total output $x = x_1 + x_2$ so as to maximize the sum of two profits $z = z_1 + z_2 = x(1 - x)$. From the necessary condition of maximum $\partial z/\partial x = 0$, we obtain the set of Pareto solutions $x = x_1 + x_2 = 1/2$ with a total profit $1/4$. Note that the losses to both players are positive as soon as their total output exceeds one. This solution is sometimes called the quasi-Pareto solution, since the number of producers (players) is only two and the actual sharing of joint profits is still indeterminate. This is however readily extended to the n player case. For example if $p = a_1 - b S$, $S = \sum_{j=1}^{n} x_j$ and total cost is $C(x_i) = c_i + c x_i$, $i=1,2,\ldots,n$ with equal marginal cost for every player, then total profit is

$$z = \sum_{i=1}^{n} z_i = a - b S^2 - A$$

where $A = \Sigma c_i$, $a = a_1 - c > 0$.

The profit-maximizing solutions are

$$S = S^* = a/2b, \quad x_i = x_i^* = S^*/n = \frac{a}{2bn}$$

price $p = a_1 - bS^* = (a_1 + c)/2$. Note that this solution is Pareto optimal from the standpoint of producers only i.e. it is a collusive or cartel like arrangement to benefit the producers. This is not Pareto optimal for the consumers, since they are paying a price $p = (a_1 + c)/2$ higher than the marginal cost c per unit of output.

(4) **Quasi-competitive solution**

The competitive solution (or quasi-competitive solution if n is small) is obtained for the demand $p = a_1 - bS$, $S = \sum_{j=1}^{n} x_j$ and total cost function $C(x_i) = c_i + c x_i$, $i=1,2,\ldots,n$ by assuming that each producer takes the price p as given and equates price to marginal cost $p = a_1 - bS = c$, thus yielding $S = S^* = a/b$ where $a = a_1 - c$ is assumed positive and $x_i = x_i^* = a/(nb)$. It is clear that compared to the collusive solution (Pareto solution for the producers only), aggregate output is now double and price much lower. Each producer is now producing double the output of a collusive producer in cartel and hence the society benefits overall. From the viewpoint of society as a whole i.e. its consumers, the quasi-competitive solution is the Pareto solution. Indeed in the theory of general equilibrium this rather than the collusive arrangement is called the Pareto equilibrium solution. The Pareto solution viewed as a competitive solution has two most important properties: it is cooperative and it is efficient. The first implies that each producer plays an atomistic role as a price taker without any concern for the other players and their outputs. This is what Cournot emphasized i.e. in his model if the number of players n gets very large, its solution would converge to the competitive one. The second property is that no reallocation of output between n producers can increase the aggregate output or reduce the price any further from the competitive level.

(5) Market-shares solution

The two players may avoid conflict and noncooperative behavior by maintaining or agreeing to maintain a desired market share. Hence if the market share of the first producer is $k = x_1/(x_1 + x_2)$, then the other producer has the share $(1-k)$. Given k and 1-k, each producer maximizes his own profit. In the mineral spring example with zero costs, this yields the solution $x_1 = k/2$, $x_2 = (1-k)/2$ with their respective profits $z_1 = (2k-k^2)/4$, $z_2 = (2-k)(1-k)/4$. This solution has two limitations. One is that it fails to explain the basis of the market share agreement and by anti-trust laws such agreements if not tacit may even be illegal and hence unimplementable. Secondly, this has an ad hoc character; if anything a bargaining scheme may have to supplement it.

(6) Discrete strategy solution

So far we have considered strategies or outputs of the two players which can vary continuously within their respective strategy sets. Now we consider discrete strategies only. Suppose for the demand function $p = a_1 - b(x_1 + x_2)$ and total costs $C(x_i) = cx_i$, (i=1,2) we consider only two strategies for each player x_1 = either u/b or 0 and x_2 = either u/b or 0 where $u = a_1 - c > 0$. These are the two extreme points of the strategy sets $X_1 = \{x_1: 0 \leq x_1 \leq u/b\}$ and $X_2 = \{x_2: 0 \leq x_2 \leq u/b\}$ for the two players, then the payoff matrix becomes

		player 2 actions	
		1	2
player 1	1	$(-u^2/b, -u^2/b)$	(0,0)
actions	2	(0,0)	(0,0)

where (z_1, z_2) denotes the payoffs to the first and second player respectively. If the best of the worst criterion (i.e. maximin criterion) is followed by each player, then both players should choose action 2 i.e produce nothing. By this they can avoid the losses of u^2/b if each produces an amount of u/b. Thus the threat of the rival that he can produce at his maximum i.e. $x_2 = u/b$ may force the producer 1 not to produce at all.

(7) <u>Nash bargaining solution</u>

The threat to be realistic must of course need be backed up by reduced costs. To show its implication assume that the producer 2 is Big and producer 1 is Small with marginal costs $c_2 < c_1$. First, Mr. Small was operating in the market with a cost function $C(x_1) = c_1 x_1$; then Mr. Big entered the market with a cost $C(x_2) = c_2 x_2$ and decided to come to an agreement rather than engaging in a costly price war. The terms of the agreement are that Big buys out Small and the question is how much price Big should pay. We can answer this question by a game-theoretical model.

The maximum profit $z_2(x_2)$ for Big when he operates alone as a monopolist is obtained by maximizing $z_2(x_2) = px_2 - c_2 x_2 = (a_1 - bx_2)x_2 - c_2 x_2$ with respect to x_2. Hence we obtain from $\partial z_2/\partial x_2 = 0$ the optimal quantity $x_2^* = (a_1 - c_2)/2b$ and the associated maximum profit $z_2^* = (a_1 - c_2)^2/2b$. This sum $s = z_2^*$ has to be divided between the two suppliers if an agreement has to be reached. But before we consider a rule of division we have to assess the consequence of non-agreement or, the threat situation. This is in accordance with Nash's theory. In case of nonagreement, the individual profits are $z_i(x_i) = [a_1 - b(x_1 + x_2)]x_i - c_i x_i$ and the optimal threats of the two players are determined in Nash theory as the outcome of a two-person zero-sum game, in which Small seeks to maximize $[z_1(x_1) - z_2(x_2)]$, whereas Big seeks to minimize this difference. But

$$z_1(x_1) - z_2(x_2) = [(a_1 - bx_1)x_1 - c_1 x_1]$$
$$- [(a_1 - bx_2)x_2 - c_2 x_2]$$

since the term $bx_1 x_2$ cancels out

hence

$$\begin{aligned}
\text{Max Min} &\ [z_1(x_1) - z_2(x_2)] \\
x_1 \quad x_2 & \\
&= \text{Max} [(a_1 - bx_1)x_1 - c_1 x_1] - \text{Max} [(a_1 - bx_2)x_2 - c_2 x_2] \\
&\quad x_1 \qquad\qquad\qquad\qquad\qquad\qquad x_2 \\
&= z_1^*(x_1^*) - z_2^*(x_2^*)
\end{aligned}$$

where $z_i^*(x_i^*)$ is the monopoly profits of each producer i=1,2. Thus the value of the game is $[z_1^*(x_1^*) - z_2^*(x_2^*)]$ for each player and the optimal threats consist of producing as if the other supplier does not exist.

A Nash equilibrium solution of the above problem consists of two parts for each player. The first part is the disagreement payoff or, the threat point, also called status quo. The second part specifies a rule of division of excess profits i.e. the difference of maximum profits resulting from their cooperation or agreement and that due to disagreement. If both players are equally rational, then equity demands that excess profits be equally divided. In the above example the excess profits is $(s - z_1^* - z_2^*)$, where $s = z_2^*$ is the maximum profits that may result from agreement. Hence the Nash equilibrium stipulates the following division rule for profits

$$\text{Small (player 1): } z_1^* + (s - z_1^* - z_2^*)/2 = z_1^*/2$$
$$\text{Big (player 2): } z_2^* + (s - z_1^* - z_2^*)/2 = z_2^* - z_1^*/2 \quad (3.3)$$

where $z_i^* = z_i^*(x_i^*)$ for $i=1,2$. Thus if we use the following parameter values $a_1 = 1.5$, $b = 1$, $c_1 = 0.9$, $c_2 = 0.5$ then the monopoly profits are $z_1^* = (a_1 - c_1)^2/2b = 0.18$, $z_2^* = (a_1 - c_2)^2/2b = 0.5$ and hence Small gets $z_1^*/2 = 0.09$ by the agreement to cooperate and sell his firm, whereas Big gets $z_2^* - 0.5z_1^* = 0.41$ in the Nash equilibrium solution.

Note that the Nash-bargaining solution (3.3) can be easily generalized to many strategies for each player and also many players. For the first case, let t_m and t_n be threats chosen by players 1 and 2 repsectively from their sets M and N respectively. Then the sharing of profits according to Nash bargaining game would be

and
$$t_m + (s - t_m - t_n)/2 \text{ for player 1}$$
$$t_n + (s - t_m - t_n)/2 \text{ for player 2.} \quad (3.4)$$

It can also be evaluated as

and
$$s/2 + (t_m - t_n)/2 \text{ for player 1}$$
$$s/2 + (t_n - t_m)/2 \text{ for player 2.}$$

To increase his payoff palyer 1 would seek to maximize the difference $(t_n - t_m)$. The position of the status quo or the threat point therefore depends on the outcome of a game in which the effective payoff function of player 1 is $(t_m - t_n)$ and that of player 2 is $(t_n - t_m)$. By adding the two payoffs we get a zero-sum game.

For the case of n-person bargaining game G assume that the payoff of player i is measured by the transferable utility u_i ($i=1,2,\ldots,n$), where it is assumed that this utility is cardinal (e.g. if it is profits in dollars then it is necessarily cardinal) and it admits interpersonal comparison. The payoff space R is denoted by

$$R = \{u \mid A \leq \sum_{i \in N} u_i \leq B\} \quad (3.5)$$

where N is the set of n players and A, B are positive constants (B > A). Let the threat point or, status quo be denoted by $t = (t_1, t_2, \ldots, t_n)$. Then the Nash solution of this game G is characterized by the payoff vector $u = u^* = (u_i^*)$, where u_i^* satisfies the following rule of division of utility (or profits)

$$u_i = t_i + \frac{1}{n}(B - \sum_{j \in N} t_j)$$

$$= \frac{1}{n} B + \frac{n-1}{n} t_i + \frac{1}{n} \sum_{\substack{j \in N \\ j \neq i}} t_j. \qquad \text{all } i=1,2,\ldots,n \\ j=1,2,\ldots,n$$

This Nash bargaining solution maximizes the product $J = (u_1 - t_1)(u_2 - t_2)\ldots(u_n - t_n)$, since all players receive the same net payoff

$$u_i - t_i = u_j - t_j = \frac{1}{n}(B - \sum_{k \in N} t_k).$$

To prove this we may transform the n-person game G into sets of two-person bargaining game and then define the n-person bargaining solution or equilibrium by the condition that there should be bilateral bargaining equilibrium between any two players i and j. Nash provided an axiomatic basis for his concept of equilibrium.

Note that the Cournot solution is generalized by the bargaining notion of Nash, since the former now can be partly cooperative and partly noncooperative. Also, the case of differentiated products sold by different suppliers (e.g. differentiation due to cost or to brand loyalty by consumers can be handled).

5.2.3 Bi-matrix Non-cooperative Game

Wage negotiations in labor market provide an interesting field of application of game theory besides the market game in duopoly or oligopoly. The labor union (player 1) wants high wages while management (player 2) wishes to give as small a wage increase as necessary. The game is clearly zero-sum, since if a_{ij} is the union's wage increase for player 1 choosing strategy i and player 2 choosing strategy j, then $-a_{ij}$ is the loss by the management. For example the union's payoff matrix for wage increase expressed in cents per hour may be

		management strategy 1	2	worst	best
union	1	40	20	20	20
strategies	2	10	30	10	
	worst	40	30		
	best		30		

Clearly there is no pure strategy saddle point equilibrium. Hence we introduce the mixed strategies $(x_1, x_2, \ldots, x_m)$ for player 1 and $(y_1, y_2, \ldots, y_n)$ for player 2 assuming that each has a finite number of strategies (here of course m = n = 2). The expected payoff of player 1 is $\sum_{i=1}^{m} \sum_{j=1}^{n} x_i a_{ij} y_j = x^T A y$ where T is transpose and x, y are the strategy vectors of the two players. The set of all mixed strategies that player 1 and 2 can use is given by $X = \{x: x_i \geq 0, \sum_{i=1}^{m} x_i = 1,$

$i=1,2,\ldots,m\}$, and $Y = \{y: y_j \geq 0, \sum_{j=1}^{n} y_j = 1, j=1,2,\ldots,n\}$. Since the expected payoff of player 1 is player 2's expected loss, hence player 2 would certainly choose his strategy vector y so as to minimize $x^T Ay$; likewise player 1 would maximize $x^T Ay$ with respect to his own strategy vector x. For both minimization and maximization to hold simultaneously, we must have

$$\underset{x \in X}{\text{Max}} \underset{y \in Y}{\text{Min}} F(x,y) = \underset{y \in Y}{\text{Min}} \underset{x \in X}{\text{Max}} F(x,y)$$

where $F(x,y) = x^T Ay$ in this case. As a matter of fact von Neumann's Minimax Theorem proved a more general result: For any function $F(x,y)$ not necessarily of the form $x^T Ay$ which is defined on any Cartesian product space $X \times Y$ i.e. for every $x \in X$ and every $y \in Y$ the following inequality holds

$$\underset{x \in X}{\text{Max}} \underset{y \in Y}{\text{Min}} F(x,y) \leq \underset{y \in Y}{\text{Min}} \underset{x \in X}{\text{Max}} F(x,y).$$

If in addition the function $F(x,y)$ is continuous then the equality holds and Maximin = Minimax. If $F(x,y) = x^T Ay$ then the game is called a matrix game with payoff matrix A and the equilibrium pair (x^*, y^*) of strategies provides a saddle point in mixed strategies with a common value of the game v^* as follows

(i) $x^T A y^* \leq v^* \leq x^{*T} A y$
(ii) $\underset{x \in X}{\text{Max}} \underset{y \in Y}{\text{Min}} x^T Ay = \underset{y \in Y}{\text{Min}} \underset{x \in X}{\text{Max}} x^T Ay = v^*.$

The concept of an equilibrium solution (x^*, y^*) of the matrix zero-sum game A with a finite number of strategies for each player leads to a generalization in the form of a bi-matrix game when the payoffs are not zero-sum. Let $A = (a_{ij})$ and $B = (b_{ij})$ be two m by n payoff matrices, one for player 1 and the other for player 2. Player 1 solves the bilinear programming problem

$$\underset{x \in X}{\text{Max}} J_1 = x^T A y$$

while player 2 solves his maximization problem

$$\underset{y \in Y}{\text{Max}} J_2 = x^T B y.$$

In terms of this bi-matrix game (A,B) we define a pair (x^*, y^*) to be an equilibrium pair if the following two inequalities hold

$$x^T A y^* \leq x^{*T} A y^*, \text{ for all } x \in X$$
$$x^{*T} B y \leq x^{*T} B y^*, \text{ for all } y \in Y \tag{3.6}$$

There exist several algorithms for finding an equilibrium point or all the equilibrium points of a finite bi-matrix game (A,B). Like the Minimax Theorem of von Neumann, one can prove an analogous theorem for a two-person finite bi-matrix game (A,B). This theorm of bi-matrix game says that for a pair (x^*, y^*) to be an equilibrium it is necessary and sufficient that there exist real numbers u^* and v^* such that x^*, y^*, u^*, v^* satisfy the following system of inequalities

$$x^T A y - u = 0, \quad x^T B y - v = 0$$
$$A y - u e_m \leq 0, \quad B^T x - v e_n \leq 0$$
$$x \geq 0, \quad x^T e_m = 1, \quad y \geq 0, \quad y^T e_n = 1$$

where e_m and e_n are m-tuple and n-tuple column vectors with each element unity.

The bi-matrix game formulation has two distinct advantages: it can represent both cooperation and noncooperation and it can have several equilibrium pairs of strategies leaving scope for further negotiation and bargaining. Thus let C be a linear combination of the two payoff matrices $C = w_1 A + w_2 B$, where w_1 and w_2 are scalar numbers (or weights), then one can set up a joint maximization problem for finding equilibrium points as

$$\underset{x}{\text{Max}} \underset{y}{\text{Max}} \sum_{i=1}^{m} \sum_{j=1}^{n} x_i c_{ij} y_j$$

$$\text{s.t.} \quad \sum_{i=1}^{m} x_i = 1, \; x_i \geq 0 \quad i=1,2,\ldots,m$$

$$\sum_{j=1}^{n} y_j = 1, \; y_j \geq 0, \quad j=1,2,\ldots,n.$$

This is a bilinear programming problem which can be solved by quadratic programming algorithms. Note that if $w_1 = 1$, $w_2 = -1$, we have a zero sum game. But if w_1, w_2 are nonnegative weights with $w_1 + w_2 = 1$, then the dominance of player 1 may be expressed by a weight w_1 higher than 0.5, whereas an equal value $w_1 = w_2 = 0.5$ would imply no dominance by any player.

To show that a bi-matrix game has many useful applications in imperfectly competitive or duopoly markets, consider the following duopoly game, where two producers (players) sell a product at prices p_1, p_2 where the demand function is $d = a - b(p_1 + p_2)$ where a, b are positive constants and the price strategies are bounded as $A \leq p_i \leq B$, $i=1,2$ by two positive constants A, B with $B > A$. The market shares s_1, s_2 of the two producers are given by $s_1 = g - h \cdot (p_1 - p_2)$, $s_2 = 1 - s_1$; $0 \leq s_i \leq 1$, $i=1,2$ where g and h are positive constants i.e. the market share of each supplier is assumed to depend only on the difference between prices. If their goal is to maximize the total sales revenue, then the two payoff functions are F_1 for player 1 and F_2 for player 2

$$F_1 = F_1(p_1,p_2) = p_1(a - b(p_1 + p_2))[g - h \cdot (p_1 - p_2)]$$
$$F_2 = F_2(p_1,p_2) = p_2(1 - a + b(p_1 + p_2))[g - h \cdot (p_1 - p_2)].$$

If we now allow the following finite or discrete values of $p_1, p_2 = 1,2,3,4$ and the following parameter values $a = 20$, $b = 1$, $g = 0.5$, $h = 1/6$ we obtain the two payoff matrices A and B defined before in (3.6):

$$A = \begin{bmatrix} 9 & 34/3 & 40/3 & 15 \\ 17/3 & 16 & 20 & 70/3 \\ 8 & 15 & \underline{21} & 26 \\ 0 & 28/3 & 52/3 & 24 \end{bmatrix} \quad B = \begin{bmatrix} 9 & 17/3 & 8 & 0 \\ 34/3 & 16 & 15 & 28/3 \\ 40/3 & 20 & \underline{21} & 52/3 \\ 15 & 70/3 & 26 & 24 \end{bmatrix}$$

where player 1 uses rows as strategies, whereas player 2 uses the columns. There are three equilibrium points in the game underlined above, of which the pair (21,21) dominates the other two i.e. (16,16) and (9,9) by ensuring higher payoffs for both. But it is clear if the two players would cooperate and come to an agreement, then they could have obtained (24,24) by each playing the last row and last column respectively. This cooperative aspect, which is emphasized by the Pareto solution is considered by Shapley to define the solution known as the Shapley solution.

Shapley solution

The Shapley procedure is to take the status quo payoff point (t_1,t_2) as the security level and then apply the Nash theorem i.e.

$$\underset{u \in R}{\text{Max}} \ J = (u_1 - t_1)(u_2 - t_2) \tag{3.7}$$

where the payoff space R is defined before in (3.5). Since the players seek any payoff equal to or better than the security status quo we have $u_i \geq t_i$, $i=1,2$. The above bargaining game denoted by $(R; t_1,t_2)$ does not explicitly take account of retaliatory threat possibilities but interprets (t_1,t_2) as the security status quo payoffs to be realized if the players fail to agree. The final solution of the bargaining game $(R; t_1,t_2)$ depends however not only on (t_1,t_2) but also on the Pareto optimal boundary of the payoff space R. The Pareto optimal boundary specifies the gain in payoff resulting from cooperation if it could be agreed to; it is hypothetical in case there is no cooperation. Consider for example the bi-matrix game

	y_1	y_2
x_1:	(-3,0)	(-1,-2)
x_2:	(2,1)	(1,3)

the Pareto optimal boundary in the payoff space R is the line segment joining the points (1,3) and (2,1), where the payoffs u_1, u_2 of the two players satisfy $u_2 = -2u_1 + 5$ for $1 \leq u_1 \leq 2$. To check $u_2 = 1$ for $u_1 = 2$ in the point (2,1) and $u_2 = 3$ for $u_1 = 1$ in the point (1,3). For any u_1 not belonging to the interval $1 \leq u_1 \leq 2$ we have negative payoff of player 1 and hence player 1 has no incentive to co-operate. Clearly the status quo or, best of the worst payoff for both players yield $t_1 = 1$, $t_2 = 0$ as the security status-quo point. Hence we maximize the Nash product $J = (u_1 - t_1)(u_2 - t_2) = (u_1 - 1)u_2$ on the Pareto optimal line $u_2 = -2u_1 + 5$ subject to $1 \leq u_1 \leq 2$. This finally yields the Shapley solution as $u_1^* = 7/4$, $u_2^* = 3/2$.

5.2.4 Differential Games over Time

Differential games are the dynamic analogues of ordinary static games i.e. the payoff functions are now intertemporal and the pure strategies for example are not points but time paths chosen to optimize an intertemporal payoff function. For the Cournot model with a dynamic demand function we have already seen an example of a two-person differential game in continuous time.

As a two-person dynamic game, a differential game offers three types of generalizations e.g. (a) as in optimal control theory it compares and contrasts the myopic and long run optimal policy for each player, (b) shows the stability over time or, otherwise of any solution or time-paths acceptable as reasonable in terms of game theory e.g. Cournot-Nash solution, and (c) through transversality conditions on the optimal trajectories it can analyze the impact of increasing or decreasing such variables as the length of the planning horizon and the exogenous rate of discount. Most important of all, the availability of equal or unequal information to the two players may be assessed in terms of its impact on the optimal time profile of the dynamic strategies chosen by the two players.

Macroeconomic applications of dynamic game theory are quite recent e.g. the recent models of "rational expectations" incorporate the notion of equilibrium of game theory with the government as one player and all other economic agents in the private sector being another. The forms of these dynamic games can be very easily illustrated by each player i maximizing his objective functional

$$\text{Max } J_i(u_1, u_2) = \int_0^T L_i(x(t), u_1(t), u_2(t))dt \quad i=1,2 \quad (4.1)$$

subject to the dynamic state equation

$$dx/dt = f(x(t), u_1(t), u_2(t)); \quad x(0) = x_0. \quad (4.2)$$

Here T is the length of the planning horizon, $x = x(t)$ is the state of the system at time t and $u_i = u_i(t)$ is the strategy of player i=1,2 at time t. For simplicity, we consider the state $x(t)$ and control variables $u_1(t)$, $u_2(t)$ to be scalar. If L_i is quadratic in x, u_1, u_2 and f is linear in x, u_1, u_2 then we obtain an LQG model in deterministic form except that the objective function (4.1) is defined for two players.

Various types of solutions in the game-theoretic sense may be characterized in terms of the differential game model (4.1), (4.2) for two players. First, we distinguish between the open-loop and closed-loop strategies for the players. In open-loop strategies the players announce and commit themselves to their control strategies as $u_i(t) = u_i(t,x_0)$ depending on time and the initial state x_0. Hence updating through additional information or through revision is not possible i.e. sometimes this may require perfect foresight of the future state of the system. Feedback or closed-loop strategies without memory are applied when players commit to control rules in the form $u_i(t) = u_i(t,x(t))$ depending on time and the current state of the system.

Second, we classify the various game-theoretic solutions as follows

(1) <u>Open-loop Cournot-Nash equilibrium</u>. this is a pair (u_1^*, u_2^*) of the differential game (4.1), (4.2) above if and only if

$$J_1(u_1^*, u_2^*) \geq J_1(u_1, u_2^*); \qquad u_1 \in U_1$$
$$J_2(u_1^*, u_2^*) \geq J_2(u_1^*, u_2); \qquad u_2 \in U_2$$

where U_1, U_2 are the sets of all feasible open-loop strategies u_1 and u_2 respectively.

(2) <u>Open-loop Stackelberg solution</u>. If player 2 is the follower acting on his reaction set $u_2 = R_2(u_1^o)$ given by maximizing $J_2(u_1, u_2)$ with respect to $u_2 \in U_2$ for each u_1^o of the leader, then $(\bar{u}_1, \bar{u}_2)$ constitutes an open-loop Stackelberg solution if

$$\bar{u}_1 = \max_{u_1 \in U_1} J_1(u_1, R_2(u_1))$$
$$\bar{u}_2 = R_2(\bar{u}_1)$$

(3) <u>Pareto solution</u>. This is obtained by the players agreeing to cooperate so as to maximize joint profits

$$J = w J_1(u_1, u_2) + (1-w) J_2(u_1, u_2), \qquad 0 \leq w \leq 1$$

by their choice of respective strategies, where w is an index of the bargaining strength of player 1. Here the open-loop strategies are most meaningful due to their agreement to cooperate.

(4) Nash bargaining solution. This is obtained by the players choosing their strategies u_1, u_2 so that the following product, also known as the Nash product is maximized i.e.

$$\text{Max } J = [J_1(u_1,u_2) - J_1(t_1,t_2)][J_2(u_1,u_2) - J_2(t_1,t_2)]$$
$$\text{s.t. } u_1 \geq t_1, u_2 \geq t_2; \quad u_1 \in U_1, u_2 \in U_2$$

where t_1, t_2 are the status quo strategies applied in case of no agreement (note that t_1, t_2 are strategies and not payoffs as in the static cases discussed earlier).

If we specialize the above nonlinear differential game model to the LQG form (i.e. J_i quadratic and the $f(\cdot)$ linear), then the closed-loop or feedback forms of the above solutions can be similarly specified.

Now consider a discrete time version of the linear quadratic differential game with

$$x(t+1) = x(t) + u_1(t) + u_2(t) \tag{4.3}$$
$$x(0) = x_0 \text{ given}$$
$$J_1 = 2x^2(1) + 2x^2(2) + u_1^2(0) + u_1^2(1): \text{ loss of player 1}$$
$$J_2 = x^2(1) + x^2(2) + u_2^2(0) + u_2^2(1): \text{ loss of player 2}.$$

Here $u_i(t)$ is the strategy of player i at time t and J_i is the cost (i.e. loss) to be minimized by each player i=1,2. For the open-loop strategies we assume that the information structure (IS) consists of only the initial state x_0 for both players and both stages t=1,2. With this IS, the Cournot-Nash equilibrium solution is obtained by expressing J_1, J_2 as functions of x_0, $u_1(t)$, $u_2(t)$ only

$$J_1 = 2(x_0 + u_1(0) + u_2(0))^2 + 2(x_0 + u_1(0) + u_2(0) + u_1(1) + u_2(1))^2 + u_1^2(0) + u_1^2(1) \tag{4.4}$$

$$J_2 = (x_0 + u_1(0) + u_2(0))^2 + (x_0 + u_1(0) + u_2(0) + u_1(1) + u_2(1))^2 + u_2^2(0) + u_2^2(1). \tag{4.5}$$

On setting to zero the derivative of J_i with respect to $u_i(t)$, t=0,1 we obtain four linear equations in four unknowns $u_1(t)$, $u_2(t)$, t=0,1. These yield the optimal solutions

$$u_1^*(0) = -10x_0/19, \quad u_1^*(1) = -2x_0/19$$
$$u_2^*(0) = -5x_0/19, \quad u_2^*(1) = -x_0/19 \tag{4.6}$$

In case of closed-loop IS, the information available to both players is the state $x(t)$ for t=0,1. It is assumed that the control strategies have no memory i.e. at stage 1 the system is at state $x(1)$ and at that stage control $u_j(1) = f(x(1))$, $j=1,2$ as a function of $x(1)$ only must be applied by each player. Now the choice of $u_1(1)$ and $u_2(1)$ will only influence the following part of the loss functions

$$J_1 = 2x^2(2) + u_1^2(1) = 2[x(1) + u_1(1) + u_2(1)]^2 + u_1^2(1)$$
$$J_2 = x^2(2) + u_2^2(1) = [x(1) + u_1(1) + u_2(1)]^2 + u_2^2(1).$$

Minimizing these functions with respect to $u_1(1)$ and $u_2(1)$ simultaneously we obtain

$$3u_1(1) + 2u_2(1) = -2x(1)$$
$$u_1(1) + 2u_2(1) = -x(1).$$

These yield the optimal values $u_1^*(1) = -x(1)/2$ and $u_2^*(1) = -x(1)/4$. For obtaining the optimal values $u_1^*(0)$, $u_2^*(0)$, we substitute these values for stage 1 into J_1 and J_2 as required by the dynamic programming algorithm and minimize each J_i with respect to $u_i(0)$, $i=1,2$. This yields the optimal values $u_1^*(0) = -19x_0/36$ and $u_2^*(0) = -x_0/4$. It is clear that these optimal solutions under closed-loop IS differ significantly from the open-loop IS solutions given by (4.6).

5.3 Concepts in n-person Games

The theory of n-person games for $n > 2$ has many potential applications in economics e.g. oligopoly, general equilibrium models of exchange and the allocation of joint costs to various activities. The existing theory, which is still developing is also very complicated due to two reasons. One is the difficulty of generalizing: a solution for n=3 may not hold for n=5. The second reason is that there exists a wide variety of so-called solutions. Empirical verifications or econometric tests of these alternative solutions are few and far between. Hence we present some basic concepts from the n-person game theory and analyze some illustrative applications.

5.3.1 Coalitions

In n-person games any player may form coalition with one or more players from the remaining n-1 players. The incentive to form a coalition of any size or to join one already existing depends on the extra payoff or returns he may obtain. This allocation or sharing of payoffs of a coalition is called <u>imputation</u>, if it satisfies two conditions:

(i) each player must obtain at least as much as he can get for himself without any one's help (this is called the individual rationality criterion), and

(ii) the total allocation of all payoffs to all the players combined equals the maximum amount they can obtain by forming one grand coalition of all the players (this is called the group rationality criterion).

While the individual rationality criterion is widely accepted among economists, the group rationality criterion is not, since the latter is not practiced very widely by the economic agents. Two common examples are the cartel formed by the oil producers in OPEC, which tends to break often times and the farmers in U.S. agriculture who cannot restrict their total output so as to maximize their returns and profits.

5.3.2 Core

Some imputations may satisfy a condition of group rationality much stronger than (ii) above. The set of all undominated imputations in a game is called the <u>core</u>, where the undominated imputations are defined as follows: let x and y be two imputations and let S be a coalition; we say x dominates y through S if $x_i > y_i$ for all $i \in S$ and $\sum_{i \in S} x_i \leq v(S)$ where $v(S)$ is the payoff or the amount of utility that the members of coalition S can obtain from the game, whatever the remaining players may do. Thus for an n-person game the core denoted by $C(v)$ is the set of all n-tuple vectors $x = (x_i)$ satisfying the two conditions

(a) $\sum_{i \in S} x_i \geq v(S)$ for all S in N

(b) $\sum_{i \in N} x_i = v(N)$ (5.1)

where $N = \{1,2,\ldots,n\}$ is the set of all players. However many games do not have a core because they do not fulfill the stronger form of group rationality which requires the condition (a) of (5.1).

In summary we may say that the core of a bargaining game describes the chance of total cooperation of all n players in the same coalition. This n-player coalition is denoted by N and it has the sum $v(N)$ available for distribution among n players: $x = x_i$ ($i=1,2,\ldots,n$). By definition we have $\sum_{i \in N} x_i = v(N)$. Such an allocation $x = (x_i)$ is called an imputation. What is the possibility that the n players fail to come to an agreement to cooperate? Let S be a coalition of a subset of n players which decides to act separately, this coalition will be able to acquire the sum $v(S)$. If the members of this coalition S do not obtain at least as much as in $x = (x_i)$, they would have made a wrong calculation in leaving the grand coalition N i.e. if the condition $v(S) \leq \sum_{i \in S} x_i$ is satisfied, the coalition would not separate itself from the larger group N. Thus, if this inequality holds for <u>all</u> coalitions S belonging to the set K_N of 2^n coalitions including the empty coalition ϕ, then the imputation $x = (x_i)$ secures a stable cooperation between all the n players. This imputation defines the core of the game. It can be shown that the core of the game is nonempty if and only if the objective function of the following LP model has a value zero at the optimum

$$\text{Min } J = \sum_{i \in N} x_i - v(N)$$

$$\text{s.t.} \quad \sum_{i \in S} x_i \geq v(S), \text{ for all } S \text{ in } K_N$$

$$x_i \geq 0.$$

This result is helpful in finding out a non-empty core through a linear programming computation, provided it exists.

5.3.3 Characteristic function

For any coalition S of a gven size, we have two groups of players S and N-S, where N is the set N = {1,2,...,n} of all players. The characteristic function of a coalition S denoted by v(S), where S is any proper subset of N is the amount of payoff or utility that the members can obtain from the game, whatever the remaining players may do. Hence v(S) is the maximin value (i.e. best of the worst payoff) to S of the two-person game played between S and N-S. Thus the characteristic function concentrates on the worst outcomes for each coalition. The theory of n-person games developed by von Neumann and Morgenstern used the characteristic function to define a 'solution' of the game as a set of imputations which has two characteristics as follows: (a) if $x = (x_i)$ and $y = (y_i)$ are any two imputations in the 'solutions', then neither x dominates y nor y dominates x, and (b) for every imputation not belonging to the solution set, there exists at least one imputation in the solution set which dominates it.

The characteristic function v(S) of a coalition S where S is a subset of the set N of n players has some important features. For instance, it has the <u>superadditive property</u> i.e. if S and T are two disjoint coalitions of N, then

$$v(S \cup T) \geq v(S) + v(T), \quad \text{for } S \cap T = \phi$$

where SUT denotes the union of members of sets S and T, S ∩ T is the intersection (i.e. members common to both S and T) and φ denotes the null set containing zero only. In economic terms the superadditivity property is merely the existence of increasing or constant returns to scale when new coalitions are formed by taking unions of the sets S and T. Clearly $v(\phi) = 0$, since φ is the null set.

Secondly, the characteristic function may be used to describe a constant-sum and therefore a zero-sum game between two players, S and N-S interpreted as two players and assuming the game to be finite. The Minimax Theorem of two-person game yields

$$v(S) + v(N-S) = v(N) \qquad (5.2)$$

for a constant-sum game and if v(N) is zero the game is zero-sum, provided v(S) + v(N-S) = 0 holds for all coalitions S in N.

Thirdly, a many person game is called <u>essential</u> if $v(N) > \sum_{i \in N} v(\{i\})$, otherwise it is inessential. Here $v(\{i\})$ is the payoff or utility from coaltion with one member i only. Clearly if games are not essential in this sense, then there is no incentive by anyone to form coalitions and since the n-person theory of games is in essence a study of the formation of coalitions, only essential games are of interest to this theory. For example, the concept of core of an n-person game as the set of undominated imputations only applies to essential games.

Finally, the coalitions may be used to characterize the optimality of a group choice as in Cournot equilibrium. For instance, let S be any coalition formed from the grand set N of n players and let $u_i(a)$ be the individual preference or utility of member i in S for any state $a \in A$, where A is the set of possible states, then we can define a group preference function by $U_S(a) = \sum_{i \in S} w_i u_i(a)$ which is the weighted sum of the cardinal utilities or payoffs $u_i(a)$. By a change of scale each weight w_i can be set equal to one and then $U_S(a)$ may be interpreted as the total winnings of the members of coalition S from state $a \in A$. Now we can define a state a^* to be optimal in S i.e. S-optimal with respect to the group preference function $U_S(a)$ if there exists no other distinct state b in A such that $U_S(b) > U_S(a)$.

Now consider a set W of coalitions with proper subsets S as sub-coalitions. The state a^* is defined to be W-optimal if it is optimal for all S in W. If W consists only of all one-element coalitions i.e. $W = (\{1\},\{2\},\ldots,\{i\},\ldots,\{m\})$, then the optimal state a^* is called a Cournot equilibrium based on individual rationality. If W consists of all players in N the grand coalition, then the W-optimal state is called the Pareto optimum (based on collective rationality). The most complete picture of collective or group rationality is conveyed by the concept of Edgeworth equilibrium. An Edgeworth equilibrium, also called the points on the contract curve, is nothing but a W-optimal state for the set W of all non-empty coalitions of the grand coalition N. This Edgeworth optimality expresses the concept of complete stability of a solution i.e. no coalition gains are possible by changing from an Edgeworth equilibrium point (or points).

5.4 <u>Economic Applications of Core</u>

As a first example of the core, we assume a market with one buyer (B) and two sellers (A_1, A_2) each selling one unit of a single product with the same constant marginal cost $c > 0$. The buyer B desires to buy at most one unit of the good at a price $b > 0$. Let x_i be the imputation of seller A_i (i=1,2) (i.e. the amount obtainable from the sale), then we must have $x_i \geq c$, i=1,2 otherwise there would be no sale. Furthermore, if the imputations x_1, x_2 for the sellers and y for the buyer are in the core defined in (5.1) then they must satisfy the following inequalities

$$x_i \geq c \qquad (i=1,2) \qquad (6.1)$$
$$x_i + y \geq v(A_i, B) \qquad (i=1,2) \qquad (6.2)$$
$$x_1 + x_2 + y \leq v(N) \qquad \text{(feasibility)}. \qquad (6.3)$$

But for the core, the constraint (6.3) must be an equality i.e. group rationality. To check if the equality holds, we consider for $i \neq i'$

$$v(N) = \max \{c + \max (c,b)\} \leq c + \max (c,b)$$
$$= v(A_i) + v(A_{i'}, B). \qquad (6.4)$$

Hence

$$x_i + y + x_{i'} \geq v(A_i) + v(A_{i'}, B) \geq v(N)$$

yielding an equality for (6.3). Hence

$$x_1 + x_2 + y = v(N). \qquad (6.5)$$

Two cases are now possible. If the two players do not collude then we obtain from (6.5)

$$x_2 = v(N) - (x_1 + y) \leq v(N) - v(A_1, B) = c$$

since $v(N) = v(A_2) - v(A_1,B)$ from (6.4) and by using (6.1), we get $x_2 = c$. Similarly $x_1 = c$. In the second case if the sellers form a collusion, then the two inequalities of (6.2) no longer apply and the imputations satisfy $c \leq x_i \leq b$ which are the two limits of bilateral monopoly. Additional rules of bargaining among sellers are needed to divide their potential gain from collusion.

The second example is one of allocating joint costs to several departments or purposes indexed by player i. The coalition S is a subset of the grand coalition N of all n players (or departments). The utility functions $v(S)$, $v(N)$ are now in terms of costs, so that they are denoted by $C(S)$ and $C(N)$. For games with cores, the minimum feasible cost within the core which can be assigned to each player i can be obtained by solving the following linear program for each player i

Minimize x_i
s.t. $\quad x_i \leq C(i), \quad$ all $i \in N$
$\quad \sum_{i \in S} x_i \leq C(S),$ all $S \in N$

$\quad \sum_{i \in N} x_i \geq C(N) \qquad (6.5)$

$\quad x_i \geq 0, \quad$ all $i \in N$.

Here the constraint set simply delimits the core of the game so that the minimum value of x_i is the lowest feasible cost that can be assigned to player i without violating the core conditions. Note that if the game has no core and the LP problem (6.5) is solved, we would obtain $\sum_{i \in N} x_i^* > C(N)$ at the optimal solutions. But for the core we need to satisfy $\sum_{i \in N} x_i^* = C(N)$. A way to satisfy this condition is to make the conditions on the intermediate coalitions less restrictive. Thus if no core exists we solve the modified LP problem

Minimize z

s.t. $\quad x_i \leq C(i), \quad$ all $i \in N$

$\sum_{i \in S} x_i - z\, C(S) \leq C(S) \quad$ all $S \in N$ \hfill (6.6)

$\sum_{i \in N} x_i = C(N)$ and $x_i \geq 0, \quad i \in N$.

The optimal solution to this LP problem yields the minimal value z^* which yields a nonempty core. On using this value we may now determine the minimal cost x_i^* for player i by solving the following LP problem

Minimize x_i

s.t. $\quad x_i \leq C(i), \quad$ all $i \in N$

$\sum_{i \in S} x_i \leq (1 + z^*)\, C(S), \quad$ all $S \in N$ \hfill (6.7)

$\sum_{i \in N} x_i = C(N)$ and $x_i \geq 0, \quad$ all $i \in N$.

As a numerical example of 3-player cost game, let $C(1) = 4$, $C(2) = 4$, $C(3) = 5$, $C(1,2) = 5$, $C(1,3) = 8$, $C(2,3) = 6$ and $C(1,2,3) = 10$. These characteristic function values imply the following constraints on the imputations or charges x_i

$x_1 = 4$, $2 \leq x_2 \leq 4$ and $x_3 = 5$.

But with $x_1 = 4$ and $x_3 = 5$, we obtain $x_2 = 1$ which contradicts $2 \leq x_2 \leq 4$. Hence the constraints of (6.5) are not mutually feasible i.e. core does not exist. Hence we apply the method outlined in (6.6) and obtain $z^* = 0.0526$ and the resulting core imputations solved from (6.7) are $x_1^* = 3.68$, $x_2^* = 1.58$ and $x_3^* = 4.74$.

This modified core solution also called ε-core for $\varepsilon = z^*$ has been applied in varoius cost allocation problems e.g. the problem of allocating the cost of an airport runway among several different sizes of aircraft. Two other popular notions of solution of cooperative game theory are often considered. One is the Shapley value which is based on the argument that each player should receive the expected value over all coalition formation sequences of the incremental value he brings to each coalition i.e.

$$x_i = \sum_{\substack{i \in S \\ S \in N}} \frac{(s-1)!\,(n-s)!}{n!} [v(S) - v(S - \{i\})]$$

The second notion is the concept of <u>nucleolus</u> introduced by D. Schmeidler (1969), which is the solution which maximizes the minimum savings accruing to any coalition. Again this can be found by solving at most n-1 linear programming models for an n-person game.

To give a simple example of the Shapley value, consider the cost example but the costs are now given as follows

$$C(x) = \begin{cases} 2x & \text{for } 0 \le x \le 60 \\ 3x/2 + 30 & \text{for } 60 \le x \le 80 \\ x + 70 & \text{for } x \ge 80 \end{cases}$$

where x is the amount of service and C(i) is the cost of department (player) i=1,2,3. Assume that the services required by each department are: $x_1 = 50$, $x_2 = 30$ and $x_3 = 30$; hence the costs are

C(1) = 100, C(2) = C(3) = 60, C(1,2) = C(1,3) = 150,
C(2,3) = 120 and C(1,2,3) = 180.

It is clear that the least cost way is for the three departments to jointly procure the service for 180 dollars. The Shapley value would then allocate these common costs to the three departments as follows

Dept 1: 100 - (0 + 10/6 + 10/6 + 40/3) = $83.34
Dept 2: 60 - (0 + 0/6 + 10/6 + 30/3) = $48.33
Dept 3: 60 - (0 + 0/6 + 10/6 + 30/3) = $48.33
      Total 180.00.

If the two departments 2 and 3 merged and considered as one department 23, then the Shapley value would allocate the total cost of $180.00 as:

Dept 1: 100 - (0 + 40/2) = $ 80
Dept 2: 120 - (0 + 40/2) = $100.

Thus it is clear that the Shapley value may be very sensitive to mergers, just as the core allocation may be affected very often by infeasibility or nonexistence.

Finally, we consider an application to a competitive exchange equilibrium, where there are m commodities x_i^j and n players or traders; each player i has a commodity bundle $w^i = (w_1^i, w_2^i, \ldots, w_m^i)$ for i=1,2,...,n as the initial endowment at the beginning of the trading period and each has a utility function $u^i(x_1, x_2, \ldots, x_m)$ of the bundle of goods $(x_1, x_2, \ldots, x_m)$. Let S be a coalition formed from the set of n players. If this coalition forms we assume for simplicity that they can redistribute their endowments in any way desired. Thus any redistribution or allocation is an S-tuple of commodity bundles $X(S) = (x^i)_{i \in S} = (x_1^i, x_2^i, \ldots, x_m^i)_{i \in S}$ satisfying the two conditions

$$\sum_{i \in S} x_j^i = \sum_{i \in S} w_j^i, \quad \text{all } j=1,2,\ldots,m \tag{7.1}$$

and

$$x_j^i \geq 0, \text{ all } i \in S, j=1,2,\ldots,m \tag{7.2}$$

Any such S-tuple of commodity bundles is called a <u>feasible allocation</u> for coalition S.

Now assume that a system of prices $(p_1, p_2, \ldots, p_m)$ is given in the market one for each commodity. The objective of each player i is to exchange his initial bundle (vector) w^i for a new bundle (vector) x^i so as to maximize his utility i.e.

$$\text{Max } u^i(x_1^i, x_2^i, \ldots, x_m^i) \tag{7.3}$$

$$\text{s.t.} \quad \sum_{j=1}^{m} p_j x_j^i \leq \sum_{j=1}^{m} p_j w_j^i \tag{7.4}$$

$$x_j^i \geq 0, \quad j=1,2,\ldots,m. \tag{7.5}$$

If the prices p_j are all positive, the set of constraints (7.4), (7.5) is closed, convex, bounded and nonempty, and the utility functions $u^i(\cdot)$ are all continuous, then the maximum solution of the above problem exists. Assuming this existence we define a competitive exchange equilibrium by a pair $(p, \bar{X})$ of vectors where $p = (p_1, p_2, \ldots, p_m)$ is a price vector and $\bar{X} = (x^i)_{i \in N}$ is any feasible allocation (i.e. it satisfies $\sum_{i \in N} x_j^i = \sum_{i \in N} w_j^i$, $j=1,2,\ldots,m$ and $x_j^i \geq 0$ for $i=1,2,\ldots,n$; $j=1,2,\ldots,m$ and $N = (1,2,\ldots,n))$ such that for each trader i we have the maximum utility $u^i(\hat{x}^i)$ attained i.e.

$$u^i(\hat{x}^i) = \max u^i(x^i)$$

$$\text{s.t.} \quad \sum_{j=1}^{m} p_j x_j^i \leq \sum_{j=1}^{m} p_j w_j^i$$

$$x_j^i \geq 0, \quad j=1,2,\ldots,m.$$

It can be shown that if each function u^i is continuous concave and nondecreasing in each argument, then such a competitive equilibrium pair $(p, \bar{X})$ exists. We may prove further that if (p^*, X) is any competitive equilibrium pair, then it must belong to the core of the market game i.e. the payoff $z_i = u^i(x_1^i, x_2^i, \ldots, x_m^i)$ must belong to the core. To prove the latter result, suppose the vector $z = (z_i)$ does not belong to the core. Then there must exist some nonempty coalition S in the grand coalition N and some allocation vector $\hat{z}^i = (\hat{z}_1^i, \hat{z}_2^i, \ldots, \hat{z}_m^i)$ feasible for S such that

$$u^i(\hat{z}_1^i, \ldots, \hat{z}_m^i) > z_i \text{ for all } i \in S.$$

Now the commodity bundle x^i maximizes u^i subject to the constraints (7.4) and (7.5) and so the vector $\hat{z}^i$ cannot satisfy them. But by feasibility $\hat{z}^i \geq 0$ and hence

$$\sum_{j=1}^{m} p_j^* \hat{z}_j^i > \sum_{j=1}^{m} p_j^* w_j^i.$$

On summing over all $i \in S$, we get

$$\sum_{j=1}^{m} p_j^* \left(\sum_{i \in S} \hat{z}_j^i - \sum_{i \in S} w_j^i \right) > 0$$

but this is impossible, since all prices p_j^* are nonnegative and the term in parenthesis i.e. $(\sum_{i \in S} \hat{z}_j^i - \sum_{i \in S} w_j^i)$ cannot be positive. Hence by contradiction it is proved that the vector $z = (z_i)$ of allocations is an undominated imputation. Hence it must belong to the core.

Note however that there may be several vector points like z belonging to the core. Also, any allocation z which belongs to the core is Pareto optimal but all Pareto optimal allocations need not belong to the core, since the concept of core imposes stronger conditions of group rationality.

Thus the core is a more restricted equilibrium than the Pareto equilibrium and hence the set of prices associated with the core has all the efficiency properties of a Pareto equilibrium. All points in the core are in equilibrium in the sense that no coalition has the incentive to disrupt them. Thus these points define a stable set.

PROBLEMS FIVE

1. The following payoff matrix is of player A in a constant-sum game

		B's strategy			
		1	2	3	4
	1	50	90	18	25
A's strategy	2	27	5	9	95
	3	64	30	12	20

   (a) What would be the best of the worst solution for player A and B?
   (b) If both players want to play safe, which solution would they prefer? Give reasons.

2. How would the Cournot solution be modified if the demand function is modified as $p = 1 - (x_1 + x_2) + e$, where e is a stochastic error term independent of x_1 and x_2 with the following values

value of e	probability
-0.5	0.25
0	0.50
0.5	0.25

   and each supplier maximizes the risk-adjusted profit $z_i = px_i - 0.1\ vx_i^2$ where v is the variance of the error term e? How would the optimal equilibrium proftis and outputs change if the variance v increases?

3. A manufacturer is considering three investment plans A, B and C under three economic conditions: increase, stable and decrease in overall demand. The payoff matrix is

	Economic Conditions		
Alternatives	increase	stable	decrease
A	$5,000	7,000	3,000
B	2,000	10,000	6,000
C	4,000	4,000	4,000

   Determine the best investment plan using each of the following criteria: Laplace, maximin, Hurwicz (w = 0.4) and minimax regret.

4. Determine the optimal time profie of output $x_1(t)$ of the dominant firm when it is maximizing its long run discounted profits

$$J = \int_0^\infty z_1(t) \exp(-rt)dt$$

$$z_1(t) = p\, x_1(t) - c\, x_1(t)$$

under the threat of potential entry by his rivals given as

$$dx_2/dt = k(p - p_0), \qquad p > p_0$$

where k, c are positive constants, p_0 is the competitive (limit) price assumed fixed and

$$p = p(t) = a \exp(nt) - b(x_1(t) + x_2(t)).$$

This is a modified verstion of the dynamic model of limit pricing under the threat of potential entry. Compare the steady state solution on the optimal trajectory with the single firm monopoly solution and the competitive solution.

5. Consider the game known as the prisoner's dilemma with payoffs for the two players as

$$\begin{matrix} (5,5) & (0,10) \\ (10,0) & (1,1) \end{matrix}$$

show that the only equilibrium pair for this bimatrix game is given by the second pure strategy of each player yielding payoffs (1,1). What is the Pareto solution?

6. Let $g_1(\cdot)$ and $g_2(\cdot)$ be the payoffs (i.e. costs to be minimized) of the two players $g_1 = (u_1 - 2)^2 + (u_2 - 1)^2$, $g_2 = (u_1 - 1)^2 + \frac{1}{2}(u_2 - 2)^2$ where u_i is the strategy (control) of player i ($i=1,2$) and the constraints on the strategies are $0 \leq u_1 \leq 4$ and $0 \leq u_2 \leq 4$. Show that the Pareto minimal control set is given by $P = \{(u_1, u_2): u_2 = 2/u_1, 1 \leq u_1 \leq 2\}$. Also the Nash equilibrium is $(\hat{u}_1, \hat{u}_2) = (2,2)$ and the minimax points are $(u_1^*, u_2^*) = (2,4)$ for player 1 and $(u_1^*, u_2^*) = (4,2)$ for player 2.

7. Consider the search game model with $x = (x_1, x_2, \ldots, x_n)$ and $y = (y_1, y_2, \ldots, y_n)$ as the strategies of players and 2 where $\sum_{i=1}^{n} x_i = A$, $\sum_{i=1}^{n} y_i = B$, $x_i \geq 0$,

$y_i \geq 0$, $i=1,2,\ldots,n$ and the payoff function is

$$F(x,y) = \sum_{i=1}^{n} b_i x_i \exp(-a_i y_i)$$

where all a_i, b_i are positive. Player 1 controlling x is the maximizing player, whereas player 2 controlling y is the minimizing player. Show that the pair (x^*,y^*) of optimal strategies satisfies the saddle point inequalities

$F(x^*,y) \geq F(x^*,y^*)$ for all y
$F(x,y^*) \leq F(x^*,y^*)$ for all x.

If r and s are the Lagrange multipliers associated with the constraints $\Sigma x_i = A$ and $\Sigma y_i = B$ respectively, then use the Kuhn-Tucker theory to show the following: (a) $r^* > 0$, $s^* > 0$, (b) $y_i^* > 0$ implies $x_i^* > 0$, (c) if $v_i > r^*$ both x_i^* and y_i^* are positive with $x_i^* = s^*/(r^* a_i)$, $y_i^* = 1/a_i \log(v_i/r^*)$ and (d) the value of the game is $r^* A$, where r^*, s^* are the optimal values of the two multipliers.

8. Let the payoff function of a zero-sum two-person game be $F(x,y) = (x-y)^2$ which is convex in y for each x. Player 1 controls x to maximize the payoff $F(x,y)$ whereas player 2 chooses y to minimize it. Show that player 2's optimal strategy is a pure strategy given by $y^* = 1/2$, whereas player 1's optimal strategy is a mixed strategy: choose $x = 0$ with probability 1/2 and $x = 1$ with probability 1/2. Show that the value of the game is $v = \min_y \max_x (x-y)^2 = \min_y \max (y^2, (1-y)^2) = 1/4$.

9. Consider the following two payoffs for a two-person noncooperative game where u_i is the strategy of player i ($u_i \geq 0$)

$J_1(u_1,u_2) = -u_1 u_2 + \tfrac{1}{2}u_1^2 + u_1$ (player 1's payoff)
$J_2(u_1,u_2) = \tfrac{1}{2}(u_2^2 - u_1^2) - (1 + u_1^2) u_2 - 2u_1$.

Show that the following two are the Cournot-Nash equilibrium solutions: $u_1 = 0$, $u_2 = 1$ and $u_1 = 1$, $u_2 = 2$. However there is no Stackelberg strategy or solution if player 2 is the leader and player 1 the follower.

10. For a three player game let the characteristic function values be: $v(\phi) = 0$, $v(\{i\}) = 0$ for $i=1,2,3$, $v(1,2) = 90$, $v(1,3) = 100$, $v(2,3) = 0$ and $v(1,2,3) = 100$. Show that the core consists of all 3-tuple vectors $x = (x_1, x_2, x_3)$ which satisfy

$x_1 + x_2 \geq 90$, $x_1 + x_3 \geq 100$, $x_1 + x_2 + x_3 = 100$ and $x_i \geq 0$, $i=1,2,3$.

Show that the core denoted by $C(v)$ may be expressed as $C(v) = \{a, 0, 100-a \mid 90 \leq a \leq 100\}$ where for example $x_1 = a$, $x_2 = 0$, $x_3 = 100-a$.

CHAPTER SIX
SOME RECENT APPLIED ECONOMIC MODELS

Recent trends in applied economic models have emphasized among others the following issues, their mathematical formulation and econometric verification: (1) how to model a stochastically optimal behavior under a dynamic environment influenced by uncertainty? (2) how to model the formation of expectations by economic agents and incorporate it in verifiable economic models? and (3) how to model the disequilibrium behavior among economic agents?

The first issue is frequently discussed in recent models of stochastic control applied in differential game models for oligopolistic markets e.g. models of limit pricing under the threat of entry, models of cautious control in monetary-fiscal policies or optimal management of uncertain resource systems in ecology. The second issue is often discussed under the hypothesis of rational expectations, although there exists several other aspects of this question not covered by the rational expectations (RE) hypothesis. The third issue analyzes some of the basic objections against the equilibrium assumption used in economic models and discusses some alternative formulations such as non-Walrasian market adjustments or, stochastic variations around a particular equilibrium. We present a brief introduction to some of these newer approaches.

6.1 Stochastic Control Models

Control theory models in economics have most frequently treated uncertainty in two ways: one is through risk aversion by the decision makers or the players in differential games and the other through updating of a policy with new information or through learning. We consider some recent examples to illustrate these two aspects.

One of the most widely used risk averse control strategy is to follow a cautious policy which combines for a decision-maker the payoffs from two apparently conflicting goals: good estimation and good regulation. Good estimation often requires a large spread in the instrument or control variables, since it would make the model more reliable and representative, whereas good regulatory control requires small variations in the instruments, since this would make the optimal policy fine tuned. As a scalar example consider one state variable (y_t) and one control (x_t) related by

$$y_t = \alpha + \beta x_t + \varepsilon_t; \quad t = 1, 2, \ldots \qquad (1.1)$$

where ε_t is a zero-mean random variable with constant variance σ_ε^2 and it is independent of control. Since this satisfies the requirement of a least squares (LS) model, we assume that the LS estimates $(\hat{\alpha}, \hat{\beta})$ are available from past observations of the pair $(x_t, y_t; t=1,2,\ldots,K)$. This is the estimation part. Now for the part of optimal (or good) regulation or control, we have to choose a performance measure or a criterion function. On using a one-period loss function L for example

$$L = E(y_{t+1} - y^F)^2 \qquad (1.2)$$

where E is expectation and y^F is a desired goal or target, and the dynamic model (1.1), one could easily compute the optimal control rule x_{t+1}^* say,

$$x_{t+1}^* = \hat{\beta}_t (y^F - \hat{a}_t) \{\hat{\beta}_t^2 + \sigma^2(\hat{\beta}_t)\}^{-1} \qquad (1.3)$$

where the LS estimates $\hat{\alpha}_t$, $\hat{\beta}_t$ are given the time subscript t, to indicate that it is computed over all observed data (x_s, y_s) available up to t i.e., $s = 1,2,\ldots,t$. Here $\sigma^2(\hat{\beta}_t)$ denotes the squared standard error of the LS estimate $\hat{\beta}_t$ i.e.,

$$\sigma^2(\hat{\beta}_t) = \sigma_\varepsilon^2 / \sum_{s=1}^{t} x_s^2 . \qquad (1.4)$$

Note the two major elements of caution in the optimal control rule (1.3). It has the squared standard error of $\hat{\beta}_t$ in the denominator, implying that a large standard error would generate a very low control x_{t+1}^*. This property would be much more magnified if we replace the one-period loss function by an intertemporal one e.g.,

$$L = E[\sum_{t=1}^{T} (y_{t+1} - y^F)^2]. \qquad (1.5)$$

This is because we have to know or guess the future values of control as of t and hence that of k-step forecast of $\hat{\beta}_{t+k}$ and its standard error. The second element of caution is due to the gap of $\hat{a}_t$ from the target level y^F. If the standard error of $\hat{a}_t$ which is $\sigma_\varepsilon/\sqrt{t}$ is very large, then the estimate of this gap is also very unreliable, thus implying the less reliability of the control rule x_{t+1}^*; also if the desired target rate y^F is set too high, it may require a large control, hence a large cost of control.

Moving the sample size from t (i.e., $s=1,2,\ldots,t$) to t+k say one could iteratively compute $(\hat{\alpha}_t, \hat{\beta}_t)$ and then x_{t+1}^* up to t+k and then reestimate $\hat{\alpha}_{t+k}$, $\hat{\beta}_{t+k}$ by LS methods and so on. Some Monte Carlo simulation experiments reported by recent researchers suggest that the successive estimates $(\hat{\alpha}_t, \hat{\beta})$ may be biased and sometimes heavily and the standard error of $\hat{\beta}_t$ may decrease to zero very slowly even under large sample conditions.

These two problems are likely to be magnified in case the objective function is of intertemporal form (1.5). Also for multivariable cases with state and control vectors, the two-stage process of control and estimation may involve substantial computational load.

A second example illustrates the problem of designing an optimal control in a fluctuating random environment. Economists have found such problems in fisheries management, where the decision problem is how to optimally harvest a randomly fluctuating population e.g., open-access ocean fishing. Consider a fishing population characterized by a scalar stochastic variable $X = X(t)$, whose dynamics of growth and fluctuations is specified by the following stochastic differential equation

$$dX = [f(X,t) - h(X,t)]dt + \sigma(X,t)dW$$
$$\text{for } X > 0. \tag{2.1}$$

Here $f(X,t)$ is the natural deterministic growth rate, $h(X,t)$ the rate of harvest, $\sigma(X,t)$ a measure of the intensity of noise and $W(t)$ a standard Brownian motion stochastic process with expectation $E\{dW\} = 0$ and $E\{(dW)^2\} = dt$. The profit from a harvest rate $h(X,t)$ is taken as

$$\pi(X,t,h(X,t)) = [p(t) - c(X,h,t)]h(X,t)$$

where $p(t)$ = price per unit harvest, $c(X,h,t)$ is the cost of harvesting a unit biomass when the stock size is X and the harvest rate is h. A long run expected value of discounted profits may then be set up as the objective function to be maximized

$$\max_{h(X,t)} J = E\{\int_0^T e^{-rt} \pi(X,t)dt + e^{-rT} S(X(T),T)\} \tag{2.2}$$

where $S(X(T),T)$ is the preservation value of stock at time T and the exogenous rate of discount is $r > 0$. By using the standard dynamic programming algorithm by setting

$$J(x,t) = \max_h E\{\int_0^T e^{-rs} \pi(X,s)ds + e^{-rs} S(X(T),T | X(t) = x\}$$

we obtain the recursive equation by the principle of optimality

$$0 = J_t + \max_h \{\tfrac{1}{2} \sigma^2(x,t) J_{xx} + [f(x,t) - h(x,t)] J_x + e^{-rt} \pi\} \tag{2.3}$$

where the subscripts of J denote the partial derivatives. This nonlinear differential equation (2.3) has to be solved for computing the optimal sequence of harvest. In some special cases the solution can be explicitly written in a closed form e.g., let the cost function $c(x)$ depend on x only and the harvest rate be independent of x, then one obtains from (2.3) the optimality condition

$$J_x = e^{-rt}[p(t) - c(x)]$$

from which the optimal value of h can be computed. However even in this case the optimal control rule h may take the form of a bang-bang control i.e., of the form

$$h = \begin{cases} h_{max} \\ h_{min} \end{cases}$$

which is socially very undesirable, since these controls are essentially "boom or bust" and due to extensive capitalization requirements they are very expensive. One way to smooth out these bang-bang type fluctuations is to modify the objective functional so that we have the following model

$$J(x,t) = \max_h E\{\int_t^\infty [(p - c(x))h - \beta(\frac{dh}{dt})^2] e^{-rs} ds \mid X(t) = x\}$$

and

$$dX = \{f(x) - h\}dt + \sigma(x)dW$$

where it is assumed $f(X,t) = f(X)$, $\sigma(X,t) = \sigma(X)$ and β is a parameter measuring the cost of changing effort.

A discrete time version of this smoothing policy which avoids the bang-bang type fluctuations is easier to apply. For example, let x_t be the population size in year t before the harvest z_t and $y_t = x_t - z_t$. The dynamics of the stock is given by

$$x_{t+1} = f(y_t, \varepsilon_t) \tag{3.1}$$

where ε_t are independent, identically distributed random variables and $f(\cdot)$ is the growth function. The objective function is

$$\max_{z_t} E\{\sum_{t=1}^\infty \alpha^{t-1} g(x_t, y_t)\} \tag{3.2}$$

where $g(x_t, y_t)$ is the value function for a harvest of y_t at stock level x_t and α is the discount rate. The optimal harvesting strategy that maximizes (3.2) under (3.1) may lead to harvests that fluctuate greatly. To smooth out such fluctuations we add a cost for these fluctuations to modify the objective functional as

$$\max E\left\{ \sum_{t=1}^{\infty} \alpha^{t-1} g(x_t, y_t) - \lambda |x_t - y_t - z_{t-1}| \right\}$$

where λ is a parameter measuring the cost of fluctuations in the harvest. As λ increases the optimal harvest policies tend to become smooth and converge towards a steady level of harvest.

Next we consider a third example from what is known as renewable resource models under uncertainty. We have two players in a Cournot-Nash differential game with the fishing effort $e_i(t)$ used in harvesting to be optimally determined. The objective of each player is to maximize the present value of long run profits i.e.,

$$\max_{e_i(t)} J_i = \int_0^{\infty} e^{-rt} \pi_i(x(t), e_i(t)) dt$$

where profit is

$$\pi_i = pq_i - c_i e_i = (pk_i x - c_i) e_i \tag{4.1}$$

and the other conditions are

$\dot{x} = f(x) - q_1(t) - q_2(t)$
$x(0) = x_0 > 0; \ 0 \leq e_i(t) \leq e_i^{max}; \ 0 < x(t) < \bar{x}$
$x(t) \geq 0; \ q_i(t) = k_i e_i(t) x(t)$
$e_i(t)$ = rate of fishing effort by player i
c_i = constant marginal cost of fishing effort
k_i = catchability coefficient
$p = p(t)$ = landed price of fish
$f(0) = f(\bar{x}) = 0$ and $f(x) > 0$ for $0 < x < x^{max}$, where x^{max} is the maximum carrying capacity
r = exogenous positive discount rate.

The demand function is taken to be of a linear form

$$p = a - b(q_1 + q_2). \tag{4.2}$$

A Cournot-Nash (CN) equilibrium in non-zero sum game is then given by a pair $e_1^*(t)$, $e_2^*(t)$ which satisfy the following conditions for all feasible trajectories $e_i(t)$, $0 \leq t < \infty$;

$$J_1(e_1^*(t), e_2^*(t)) \geq J_1(e_1(t), e_2^*(t)), \text{ for all } e_1(t)$$
$$J_2(e_1^*(t), e_2^*(t)) \geq J_2(e_1^*(t), e_2(t)), \text{ for all } e_2(t). \tag{4.3}$$

The optimal trajectory of the model (4.1) for each i=1,2, is characterized by Pontryagin's maximum principle as

$$\dot{\lambda}_i = \lambda_i [r - f'(x) + K] - e_i (ak_i - 2bKk_ix)$$

$$\dot{x} = f(x) - Kx; \quad K = \sum_{i=1}^{2} k_i e_i \qquad (4.4)$$

$$ak_i x - c_i - bx^2(K + k_i)k_i - \lambda_i k_i x = 0$$

$$\lim_{t \to \infty} \lambda_i e^{-rt} = 0.$$

Assuming a symmetric case i.e., $k_i = k_0$, $c_i = c_0$ and n players one could derive from (4.4) the following steady state equilibrium values $\bar{e}$, $\bar{q}$, $\bar{x}$

$$\bar{e} = \sum_{i=1}^{n} \bar{e}_i = \frac{n(ak_0\bar{x}-c_0)}{(n+1)bk_0^2\bar{x}^2} - \frac{\bar{\lambda}}{bk_0^2\bar{x}^2(n+1)}$$

$$\bar{q} = \sum_{i=1}^{n} \bar{q}_i = k_0 \bar{x}\bar{e} \qquad (4.5)$$

$$f(\bar{x})/\bar{x} = k_0 \bar{e}$$

$$\bar{\lambda} = \frac{[a-2bf(\bar{x})]f(\bar{x})}{\bar{x}(r+k_0\bar{e}-f'(\bar{x}))} .$$

If $f(x)$ is a logistic function i.e.,

$$f(x) = \alpha x (1 - \frac{x}{x^{max}})$$

then $\quad \bar{x} = x^{max} (1 - \frac{k_0\bar{e}}{\alpha}), \quad \frac{k_0\bar{e}}{\alpha} < 1$,

also $\quad \lim_{n \to \infty} \bar{e} = (ak_0\bar{x}-c_0)(bk_0^2\bar{x}^2)$

$\lim_{n \to \infty} \bar{q} = \frac{a}{b} - \frac{c}{bk_0\bar{x}}$

$\lim_{n \to \infty} \bar{p} = \frac{c}{k_0\bar{x}^o}, \quad \bar{x}^o = \lim_{n \to \infty} \bar{x}$.

It is clear that increasing n has the effect of increasing the total harvest at the equilibrium and hence decreasing the steady state stock level of the renewable resource; the effect of increasing the discount rate r is similar.

A simple way to introduce demand uncertainty in this model is to formulate the inverse demand function with an additive error

$$p = a - b \sum_{i=1}^{n} q_i + \varepsilon$$

where ε has a zero mean, constant variance σ^2 and is independent of total demand quantity. The objective function is then revised to include a term $\frac{w}{2} q_i^2 \sigma^2$ for the risk aversion of each player

$$\underset{e_i}{\text{Max }} J_i = \int_0^\infty e^{-rt} (\pi_i - \frac{w}{2} q_i^2 \sigma^2) dt$$

where var $\pi_i = q_i^2 \sigma^2$ = variance of profit, w is a nonnegative weight. The risk aversion term may also be interpreted as an attitude to smooth out fluctuations in profit. The higher the weight w or σ^2, the lower would be the fishing effort and the total harvest in equilibrium. It is clear that cooperative, Pareto and other minimax solutions could be easily incorporated in this framework. Similar models have been formulated in the current theory of limit pricing by the dominant firm under the threat of entry by its rivals, when the potential market penetration by entry is partly probabilistic.

6.2 Models of Rational Expectations

Models of rational expectations (RE) are attempts to relate the expectations of individual agents to the actual aggregate behavior of the market in equilibrium. Such models are increasingly applied in both micro and micro fields of economic theory and the recent econometric models frequently include a rational expectations postulate in analyzing the government policy affecting the private sector business and households. Put very simply the RE models build on a very simple postulate: the agents affected by government policy (e.g., monetary or fiscal policy) do not remain neutral but form their expectations about the future state in an optimizing manner. Hence the econometric models which calculate the optimal policies assuming the public expectations to be predetermined are probably wrong and the optimal policies are probably inconsistent. The correct way according to the RE theory is not to select policy variables taking expectations as given but to model the expectation formation itself as a function of proposed policy.

In microeconomic theory John Muth (1961) introduced for the first time the need to analyze the expectational variable (e.g., expected price p_t^*) in a competitive market for a nonstorable good with a fixed production lag

$$d_t = -a\, p_t \text{ (demand)}, \quad s_t = bp_t^* + e_t \text{ (supply)}$$
$$d_t = s_t \text{ (market equilibrium)}. \tag{1.1}$$

Here all variables, d_t (demand), s_t (supply), p_t (price) and p_t^* (price expected to prevail in period t) are measured as deviations from their equilibrium values, e_t is a stochastic error term representing for example yield variations due to weather. The error term e_t is unobservable at the time the suppliers make the output decision on the basis of p_t^* which is nothing but the market price at t expected on the basis of information available through the period t-1, but e_t is observed and hence known at the time demand is revealed and the market clears by the equilibrium condition $d_t = s_t$. The problem for the suppliers is: how to estimate from the information set S_{t-1} = {all information available through the period t-1} the equilibrium price equation

$$p_t = -bp_t^*/a - e_t/a \tag{1.2}$$

when the expectational variable p_t^* is not observable? Also, how can one interpret the equilibrium price in terms of (1.2), when the error e_t is interpreted as shocks to the system following a probability distribution?

The answer provided by Muth to these questions contains the basic logic of the RE theory. He argued that if the errors e_t have zero mean and have no serial correlation, one obtains from (1.2) by taking expectations (E) of both sides

$$E(p_t) = -bp_t^*/a \tag{1.3}$$

Now the suppliers' price expectations may either satisfy the following condition

$$E(p_t) = p_t^* \text{ (rationality postulate)} \tag{1.3}$$

known as the rationality postulate or not. In the latter case there would be scope for insiders to profit by predicting price much better. But such insider profits would be zero if the suppliers' expectations satisfy the rationality postulate (1.3). This is because we would have $p_t^* = 0$, since b/a is not equal to one, as a, b are arbitrary positive constants. In other words the rationality postulate (1.3) yields the result that expected price as expected by the suppliers equals the equilibrium price. Interpreted in another way, the rationality postulate provides an unbiased forecast of the market price parameter p_t^* when each individual supplier acting as a price taker uses his own forecast based on the information set S_{t-1}.

When the errors or shocks e_t do not have mean zero but are predictable from the past information, the RE postulate (1.3) becomes

$$p_t^* = -(a+b)^{-1} E(e_t). \tag{1.4}$$

The case when the shocks are serially correlated can be handled in a similar way. For instance let us assume the shocks to be a linear combination of the past history of normally and independently distributed random variables ε_t with zero mean and constant variance v i.e.

$$e_t = \sum_{i=0}^{\infty} w_i \varepsilon_{t-i}, \quad E(\varepsilon_i) = 0, \text{ all } i$$

$$E(\varepsilon_i \varepsilon_j) = \begin{cases} v, & \text{if } i = j \\ 0, & \text{otherwise} \end{cases} \tag{1.5}$$

then the RE postulate (1.3) would be modified as

$$p_t^* = (\frac{a}{b}) \sum_{j=1}^{\infty} (\frac{b}{a+b})^j p_{t-j}. \tag{1.6}$$

In other words if the suppliers use the forecasting rule (1.6), then they would have self-fulfilling expectations in the aggregate, since their decisions would generate price distributions through the market clearing condition such that their decisions would prove to be optimal with respect to the distributions so generated.

Two points are to be noted about the RE equilibrium. One is that each supplier need not have indentical expectations about the market price variable. What is required is that the agents' individual expectations should be distributed around the true expected value of the variable to be forecasted. Secondly, the equilibrium market price in the face of the uncertain shocks e_t is a stochastic equilibrium and not a deterministic equilibrium. In dynamic control theory models of the LQG variety, the RE hypothesis is equivalent to the requirement of the perfect foresight rule along the optimal trajectory.

In macro-dynamic policymaking it is argued that policymakers (e.g., the fiscal and monetary authority) could still compute "optimal policies" by means of an econometric model but they would have to follow a procedure different from those in which the expectations of the private sector are predetermined. Thus the policymakers may decide on a set of fixed rules by weighing a set of alternative rules while taking into account the expected reactions of the private sector to those rules. These fixed rules must be known to the public so that it can choose its own optimizing behavior. The RE model predicts that the stochastic equilibrium is in most cases stable. This means that the agents in the private sector would continue to revise their expectation functions when their expectations are found to be wrong. As agents learn more and more from new information, their forecasts will become better. In the limit they know the structure of the economy and they can forecast future prices exactly. Thus their expectations are self-fulfilling.

The RE equilibrium may be viewed as a Cournot-Nash equilibrium rather than a competitive equilibrium, when we view the players' expectation functions in the framework of conjectural variations.

Recently the RE hypothesis that in the aggregate the anticipated or expected price is an unbiased predictor of the actual price has been extended to what is known as the asymptotically rational expectations (ARE) hypothesis. The ARE argues that the expected price variable in the RE model should be replaced by the expected price adjusted for risk and risk aversion and not the expected price per se. Hence the competitive firms would produce the optimal output (x_t) at that point at which marginal cost is equal to the risk-adjusted price.

Assuming a quadratic cost ($cx_t^2/2$), a quadratic expected utility of profits function with constant rate r of absolute risk aversion and the price p_{t+1}^* to be a stochastic variable with mean $E_t(p_{t+1})$ and a constant variance v, we obtain the ARE hypothesis as

$$cx_1 = E_t(p_{t+1}) - r\,v\,x_t. \tag{1.7}$$

It is clear that as the term rv due to risk adjustment tends to zero, the risk-adjusted price converges to the expected price. To show the usefulness of the ARE hypothesis (1.7) in a macroeconomic world, consider the following model of inflationary expectations according to the current monetary theory. The anticipated rate of inflation π^* over the planning period is taken to be equal to the steady-state per capita rate of monetary expansion $\mu_e - n$, where n is the constant rate of growth of labor, μ is the current rate of monetary expansion, and the subscript e denotes steady-state value i.e.

$$\pi^* = \mu_e - n. \tag{1.8}$$

At every time point t the public is assumed to form their expectation of the value of the steady-state rate μ_e of monetary expansion as

$$d\mu_e/dt = c(\mu - \mu_e). \tag{1.9}$$

By combining (1.8) and (1.9) we obtain the equation for the anticipated rate of inflation adjusted for risk as

$$d\pi^*/dt = c(\mu - n - \pi^*) \tag{1.10}$$

where $\mu = \mu_t$ corresponds to the expected price, π^* corresponds to the risk-adjusted price in (1.7) and c is an index of speed of adjustment reflecting risk and risk aversion i.e. the uncertainties concerning future disturbances. The term ($\mu - n$) is the current rate of monetary expansion per capita, which will be the equilibrium rate of inflation denoted by π_e if the current rate continues. The RE hypothesis (1.10) implies that the public does not know if the current rate (μ_t - n) of monetary expansion will continue or not, hence they do not know the steady-state rate of inflation π_e and therefore they are required to form an estimate. Since there is uncertainty whether the current rate of inflation is permanent or transitory, rational behavior for the risk-averse public requires them to wait for more evidence before changing their risk-adjusted price anticipations.

The introduction of the RE hypothesis in macroeconomic theory and policy has led to a fundamental attack against and even the rejection of Keynesian tools of management of effective demand to cure unemployment in the developed capitalist economy. The two new theories which attempt to replace the Keynesian economics are called the new classical economics (NCE) and the monetarist approach (MA). Measuring the unemployment rate as a devation $u_t = U_t - U_e$ from its equilibrium value, the core of the NCE maintains that

$$u_t = \sum_{i=1}^{n} a_i u_{t-i} + e_t \qquad (2.1)$$

i.e., the current rate of unemployment depends on its lagged values and a stochastic error or shock term e_t, where the shocks e_t are statistically independent of all past values of monetary and fiscal policies and has white noise properties i.e. purely random with no structure. Thus if x_t is a vector of monetary and fiscal policies of the government at time t-1, we have the NCE hypothesis

$$E_{t-1}[e_t|x_{t-1}] = 0, \; E[e_t, e_{t-1}] = 0, \text{ all } t.$$

Hence we obtain from (2.1) the proposition

$$E_{t-1}(u_t) = \sum_{i=1}^{n} a_i u_{t-i}$$

which is sometimes called the policy ineffectiveness hypothesis i.e., on the average the deviation u_t in unemployment rate is totally independent of monetary-fiscal policies and hence the Keynesian demand management policies are totally ineffective in reducing the deviation in unemployment rate.

The basic proposition of MA is intermediate between the polar views of Keynesianism and the NCE. The MA maintains that inflation is mainly a monetary phenomenon and the past rates of growth of money stock are the only systematic factors determining the rate of inflation. Thus any change in government expenditure policy for demand management must be associated with a change in the rate of monetary expansion to have a substantial impact on the rate of inflation. The MA hypothesis can be stated as

$$\pi_t = a_0 + \sum_{i=i}^{n} a_i \mu_{t-i} + \varepsilon_t, \quad \sum_{i=1}^{n} a_i = 1$$

where ε_t is a random error term with zero expectation. This equation says that the rate of price inflation (π_t) is a weighted average of the past rates of monetary expansion (μ_{t-1}). Empirically such a relation fits very well by econometric tests for the U.S. monetary history from 1957 through 1979.

To conclude this section we may mention however that the debate over the RE hypothesis is not over yet; more empirical evidence and econometric testing are certainly needed to resolve the debate. We must note however that increased attention is now paid to the expectational variables in the explanation of wage-price changes in current econometric models. For example the Phillips' model of wage-inflation also known as the Phillips' curve which postulates in the short run a trade-off between wage changes ($\dot{w}_t$) and the level of unemployment (U_t) is now modified as $\dot{w}_t = f(U_t^{-1}, \dot{p}_t^e)$ where dot over a variable denotes the rate of change and p^e denotes the price expected to prevail in future. Assuming a linear version with an additive error term e_t with zero mean we obtain

$$\dot{w}_t = a_0 + a_1 U_t^{-1} + a_2 \dot{p}_t^e + e_t. \qquad (2.2)$$

But since the expectational variable $\dot{p}_t^e$ (which contains the public's view about future price inflation) is not observable, several models of expectation formation are attempted and then the goodness of econometric fit against observed empirical data is allowed to choose the most appropriate model of expectation. Thus the following expecation models are applicable to the linear form (2.2)

(i) Static expectations: $\dot{p}_t^e = \dot{p}_t$ whence (2.2) becomes $\dot{w}_t = a_0 + a_1 U_t^{-1} + a_2 \dot{p}_t + e_t$.
(ii) Extrapolative expectations: $\dot{p}_t^e = \dot{p}_t + k(\dot{p}_t - \dot{p}_{t-1})$ where k can be positive (past trends to continue), negative (past trends expected to be reversed) or zero.
(iii) Adaptive expectations: $\dot{p}_t^e = \dot{p}_{t-1}^e + b(\dot{p}_t - \dot{p}_{t-1}^e)$ with $0 \le b \le 1$, which yield

$$\dot{p}_t^e = b \sum_{i=0}^{\infty} (1-b)^i \dot{p}_{t-i} .$$

(iv) Dummy variables: these are introduced as proxy variables for upward ($\dot{p}_t^e > 0$), downward ($\dot{p}_t^e < 0$) and no change ($\dot{p}_t^e = 0$) in public's view about future price inflation.

6.3 Disequilibrium Systems

From a practical applied viewpoint the models which analyze disequilibrium per se and not as deviations from equilibrium are important for three reasons. One is that trading in many markets is subject to some limits on price and the disequilibrium occurs when these limits are violated. Thus there are minimum wage laws in labor markets and trading limits in commodity markets. Secondly, the Walrasian price adjustment process, also called the successive <u>tâtonnement</u> process which is basically dependent on the excess demand postulate (i.e., a positive (negative) excess demand tends to raise (reduce) the price) is very artificial and unrealistic, if there is no auctioneer or market manager to implement the above behavioral rule. If the agents (i.e. traders) could trade out of equilibrium, the excess demand functions would change at different moments and hence would not follow the <u>tâtonnement</u> process. Thirdly, a stochastic equilibrium may have characteristics of stability which are very different from the deterministic equilibrium. The rational expectations models have stressed one such difference but there are others.

Consider first the disequilibrium model of the first type, which has two varieties e.g., the rationing models and trading models. For the one good case the rationing model appears with a minimum condition

$$\text{demand:} \quad d_t = a_0 - a_1 p_t + a_2 z_{1t} + e_{1t} \tag{3.1}$$
$$\text{supply:} \quad s_t = b_0 + b_1 p_t + b_2 z_{2t} + e_{2t} \tag{3.2}$$
$$\text{transactions:} \quad x = \min(d_t, s_t). \tag{3.3}$$

Here p_t is the market price at time t, z_1, z_2 are the two exogenous variables, e_{it} are error terms and x is the actual volume of market transactions in the market. Note that if the price is either rigidly fixed or set exogenously, the minimum of demand and supply i.e. $x = \min(d_t, s_t)$ does not necessarily determine the actual amount of transactions in case of disequilibrium, as either sellers or buyers may accept quantities that are higher than the supply or demand according to their ex ante intentions. Rationing may also determine the actual amount of demand and supply i.e. in case of excess demand the available supply is rationed out to demanders and in case of supply the available demand is rationed out among the suppliers. By trading models we refer to those models where $x = 0$ if d_t is not equal to s_t. This would be the case when the price equating demand and supply falls outside the admissible range and no transactions occur.

The above disequilibrium model may be compared with the equilibrium version, where (3.1), (3.2) remain the same but (3.3) is replaced by a competitive Walrasian adjustment mechanism in discrete-time version i.e.

$$p_t - p_{t-1} = k(d_t - s_t), \quad k > 0. \tag{3.4}$$

Let us now introduce an unobservable variable p_t^* which is the price at which the market would clear if it could. In other words, given the current values of the exogenous variables z_1, z_2 the price p_t^* is defined by the condition $d_t = s_t$. This yields

$$p_t^* = (a_1 + b_1)^{-1} [(a_2 z_{1t} - b_2 z_{1t}) + (e_{1t} - e_{2t})].$$

Evidently if the market is to be continuously in a state of equilibrium we must have $p_t = p_t^*$ for all t. Thus we may postulate a more general adjustment mechanism by replacing (3.4) by the following

$$p_t = c p_{t-1} + (1-c) p_t^*. \qquad 0 \leq c \leq 1 \qquad (3.5)$$

Thus for the limiting case of instantaneous market equilibrium we put c=0; the other polar case of no response corresponds to c=1. The parameter c may thus be used as a measure of the drag in market clearing. It is clear that p_t is above (below) p_t^* according as d_t is less (greater) than s_t, so that the sign of $(p_t - p_t^*)$ may be used to discriminate between periods of excess supply or excess demand.

The key element in the disequilibrium market model given by (3.1) through (3.3) is the fact that the price term p_t is either rigid or exogenously fixed. If we allow however the system where the price movements, while not always clearing the market in every period are nonetheless responsive to forces of supply and demand, then we obtain a simultaneous system defined by (3.1) through (3.3) and (3.5). Problems of econometric estimation of such models have attracted much recent attention from modern researchers and their applications in various markets have been explored.

For the second type of disequilibrium we may recall that the Walrasian price adjustment process in the n-commodity case is specified by the system of n differential equations

$$dp_j(t)/dt = x_j(p_1(t), p_2(t), \ldots, p_n(t)) - \bar{x}_j \qquad j=1,2,\ldots,n \qquad (4.1)$$

where $f_j(p) = x_j(p(t)) - \bar{x}_j$ in the excess demand function for commodity j and $\bar{x}_j$ is the initial stock or endowment in good j. The Walrasian equilibrium is then specified by a price vector $\hat{p}$ for which $dp_j/dt = 0$ for all j i.e., $f_j(\hat{p}) = 0$, all $j=1,2,\ldots,n$.

Now we consider an exchange economy with K traders k=1,2,...,K where any trader k holds a vector $\bar{x}_k$ of the n commodities, i.e. $\bar{x}_k = (\bar{x}_{k1}, \bar{x}_{k2}, \ldots, \bar{x}_{kn})$. The adjustment process now involves both the price vector p(t) and the commodity vectors $\bar{x}_1, \bar{x}_2, \ldots, \bar{x}_k$ as the traders can trade in the intermediate stages of disequilibrium in the non-tatonnement process. Hence the adjustment equations of (4.1) are now replaced by two systems of equations

$$\frac{dp_j(t)}{dt} = \sum_{k=1}^{K} x_{kj}(p_1,\ldots,p_n;\bar{x}_1,\ldots,\bar{x}_K) - \sum_{k=1}^{K} \bar{x}_{kj}$$

$$j=1,2,\ldots,n; \quad k=1,2,\ldots,K \tag{4.2}$$

and

$$\frac{d\bar{x}_{kj}}{dt} = h_{kj}(p_1,\ldots,p_n;\bar{x}_1,\ldots,\bar{x}_K) \quad j=1,2,\ldots,n; \quad k=1,2,\ldots,K \tag{4.3}$$

where $p_j = p_j(t)$, $\bar{x}_k = \bar{x}_k(t)$, $\bar{x}_{kj} = \bar{x}_{kj}(t)$ and the functions $h_{kj}(\cdot)$ denote like the excess demand functions the transaction rules followed by the traders in changing their goods vectors. The equilibrium of the non-tatonnement process is now denoted by the vectors $(p^*,\bar{x}_1^*,\bar{x}_2^*,\ldots,\bar{x}_K^*)$ if they satisfy the condition

$$\sum_{k=1}^{K} x_{kj}(p^*,X^*) = \sum_{k=1}^{K} \bar{x}_{kj}(0), \quad j=1,2,\ldots,n$$

where $\bar{X}^* = (\bar{x}_1^*,\bar{x}_2^*,\ldots,\bar{x}_K^*)$. If the vectors $(p(t),\bar{X}(t))$ defined by (4.2) and (4.3) converge to $(p^*,\hat{X}^*)$ as t goes to infinity, we may say that the non-tatonnement process is stable.

The Edgeworth process known as the contract curve which we described in our discussion of game theory provides an example of a non-Walrasian trading adjustment, where each player (trader) continues to trade as long as it increases his or her satisfaction (utility or profit). Another recent example is provided by the Hahn-Negishi exchange process which has the following two aspects: (i) in equilibrium if it holds that $\sum_{k=1}^{K} x_{kj}(t) = \sum_{k=1}^{K} \bar{x}_{kj}(t)$, then for each commodity $j=1,2,\ldots,n$ it holds that $x_{kj}(t) = \bar{x}_{kj}(t)$ for all $k=1,2,\ldots,K$; and (ii) in disequilibrium if $x_{kj}(t) \neq \bar{x}_{kj}(t)$, then it holds that $\text{sign}[x_{kj}(t) - \bar{x}_{kj}(t)] = \text{sign}[\sum_{k=1}^{K} x_{kj}(t) - \sum_{k=1}^{K} \bar{x}_{kj}(t)]$ for all $k=1,2,\ldots,K$ and $j=1,2,\ldots,n$. The condition (i) says that if for any good j the excess demand or supply is zero, then each buyer or seller has achieved his or her satisfaction. The condition (ii) says that if there is an excess demand for j i.e. $\sum_k (x_{kj} - \bar{x}_{kj}) > 0$ then all sellers can sell and if there is an excess supply i.e. $\sum_k (x_{kj} - \bar{x}_{kj}) < 0$, then all buyers can buy. It is clear that this exchange process does not use the gross substitutability assumption of the Walrasian adjustment mechanism and hence its stability proved by Hahn and Negishi is more general.

Recent researchers have attempted to incorporate nonzero conjectural variations of the Cournot model as strategies of the players who trade to achieve a conjectural equilibrium in a dynamic Cournot-Nash framework.

The third aspect of disequilibrium postulates that a stochastic equilibrium is more general than a deterministic equilibrium and hence the instability of the latter does not imply the instability of the former. As an example consider a scalar stochastic linear differential equation for the state variable $x = x(t)$ which may represent the capital stock of an economy

$$dx = ax\,dt + bx\,dw. \tag{5.1}$$

Here the initial value x_0 of x is assumed given, a and b are positive constants and the component $w = w(t)$ of the differential term dw is a zero-mean Gaussian process with stationary independent increments, also known as the Brownian motion process. If we ignore the stochastic part i.e., $dw = 0$, then the solution of the deterministic part is not stable since $x(t)$ tends to infinity as t goes to infinity. But the complete stochastic model has the solution

$$x(t) = x_0 \exp[(a - b^2/2)t + b\,w(t)] \tag{5.2}$$

which is stable if $b^2 > 2a$.

Recent research has been very active in this field of stochastic equilibrium and disequilibrium and their stability characteristics. This is particularly so in dynamic models of stochastic optimization e.g., stochastic differential games in a noncooperative framework.

SELECTED READING LIST

A. Intermediate level

Chiang, A.C. Fundamental Methods of Mathematical Economics. New York: McGraw-Hill, Third edition, 1984.

Dixit, A.K. Optimization in Economic Theory. London: Oxford University Press, 1976.

Franklin, J. Methods of Mathematical Economics: Linear and Nonlinear Programming, Fixed Point Theorems. New York: Springer-Verlag, 1980.

Henderson, J.M. and R.E. Quandt. Microeconomic Theory: A Mathematical Approach. New York: McGraw-Hill, Third edition, 1980.

B. Advanced Level

Arrow, K.J. and M.D. Intriligator eds. Handbook of Mathematical Economics. Amsterdam: North Holland, 1981, 2 vols. (selected parts).

Cornwall, R.R. Introduction to the Use of General Equilibrium Models. Amsterdam: North Holland, 1984.

Murata, Y. Mathematics for Stability and Optimization of Economic Systems. New York: Academic Press, 1977.

Takayama, A. Mathematical Economics. New York: Cambridge University Press, 1983.

INDEX

Absolute value	7
Accelerator	124
Activity analysis	50
Adam Smith's theory	39
Adaptive control	170
Adaptive expectations	116, 226
Adjoint variable	148, 153
Aggregative economy	123
Amplitude	108
Arrow, K.J.	10, 155
Arrow-Karlin model	180
Asymptote	127, 222
Balanced growth	129
Bang-bang control	165
Bargaining:	
game	198
solution	206
Basic feasible solution	43
Bayes criterion	188
Bellman's principle	171
Bernoulli equation	129
Bilinear game	201
Bimatrix game	200
Bordered Hessian	80
Boundary condition	159
Bounded set	75
Brownian motion	221
Bruno model of optimal growth	166
Calculus of variations	140
Canonical equations	153
Capital per worker	128
CES production function	36
Chain rule	20
Characteristic:	
equation	32, 93
function	209
root	94
vector	32, 103

INDEX

Characteristic roots:
    of difference equations   97
    of differential equations   110
    and dynamic stability   111, 133
    of matrix   94
    of mixed equations   133
Closed loop control   205
Closed set   75
Coalition   207
Cobb-Douglas production function   129
Cobweb model   90
Coefficient matrix   30
Coefficients of production   33
Cofactor   15
Commodity space   75
Compact set   75
Comparative statics:
    of market models   28, 116
    of national income models   123
Competitive economy   130, 141
Competitive equilibrium and Pareto optimality   141, 232
Complementary slackness   57
Complex numbers   95
Concave functions   10
    criteria for checking   24
    concave programming   39, 64
Conjectural variations   119
    in Cournot-Nash games   193
Constant returns to scale   128
Constant sum game   209
Constraint qualifications (CQ)   69
    application of   70
Consumption function   123
Continuity:
    in functions   19
    in series expansion   21
Contract curve   234
    in Edgeworth market game   210
Control problems   163
Convex function   24

and convex set	75
criteria for checking	24
in nonlinear programming	64
Cooperative game	196, 200
Core:	
allocation	208
and competitive equilibrium	210
and n-person game	209
and Pareto efficiency	210
Corner solution	39
Cournot solution	193
Cramer's rule	16
Critique of optimization	82
Degeneracy	47
Degree:	
of difference equation	97
of differential equation	110
Demand:	
elasticity	37
excess	130
De Moivre's theorem	95
Dependence:	
linear	28-29
Derivative rules	19-21
Derived demand function	155
Determinant	14-16
Determinantal equation	32
Determinantal test:	
for relative extremum	79-81
for nonsingularity	15
Differential equation	104-111
autonomous	130
general solution of	110
particular solution of	111
Differential games	204-206
coorperative	204
noncooperative	205
two-person nonzero-sum	205
Differentiation rules	19-21
Discounting	146
Discrete time	86, 88-90

Discriminant	92, 105
Diagonal matrix	28-30
Difference equation	88-98
general solution of	97
particular solution of	98
Disequilibrium systems	232-235
Distance:	
Euclidean	75
Domain	24
Dominant root	33-35
D-partition method	133
Duality theorems	54-60
and saddle points	68-72
Dual problem	54
economic interpretation of	58
Duffing equation	139
Duopoly model	119
and Cournot solution	193
and Stackelberg solution	195
Dynamics of:	
inflation and unemployment	124
input-output models	115
market price	116
Solow model	128
Walrasian general equilibrium	130
Dynamic programming	171-181
applications of	172
Dynamic stability of equilibrium	130
with continuous time	105
with discrete time	92
and Routh Theorem	111
and Schur Theorem	97
Econometrics	220-232
Economy-wide model	34, 123
Edgeworth contract curve	234
Edgeworth game	210
Efficiency:	
in core	210
set-theoretic concepts of	75
Eigenvalue	32
Eigenvector	32

Elasticity of demand	37
Equations of motion	153
Equation system:	
consistency	30
dynamic	86-111
homogeneous	31
linear	30
nonlinear	121,127
Equilibrium	116,123
dynamic stability of	97,110
intertemporal	148,158
stochastic	220
Equilibrium point of a game	190-198,200
Euclidean distance	75
Euler-Lagrange condition	154
Euler's Theorem	36
Excess demand	130-132
and price adjustment	232-235
Expectations:	
adaptive	116
extrapolative	231
static	231
Extreme point	42
Extremum	39-79
absolute versus relative	64
in relation to concavity	24
Feasible region	39
Final demand	33
First-order condition	24
Fisheries management	121
Fluctuation	92,105
French planning	59
Frobenius Theorem	33
application of	34
Function	8-10
concave versus convex	24
continuous	19-21
homogeneous of degree r	36
implicit	25
quasiconcave	72

Function of n variables	21
linear form	28
quadratic form	24
Functional	141-170
Fundamental equation of	
neoclassical growth	128-129
Game	182-185
against nature	187
constant sum	207
cooperative	196
core	208
Cournot-Nash equilibrium	193
differential	204
equilibrium point	200-202
in characteristic function	209
in extensive form	182
in normal form	182
imputation	207
mixed strategy	190
Nash cooperative solution	200
noncooperative	198-204
nonzero-sum	200
Pareto solution	196
payoff function	187, 200
payoff matrix	187
pure strategy	187-190
saddle point	54-64
Shapley value	203
status quo point	198-203
threat payoff	203
tree	182
two-person nonzero-sum	190-198
two-person zero-sum	187-190
zero-sum	187
Gaussian process	220-225
General equilibrium	130
input-output model	33-35
excess demand approach	132
Walrasian tâtonnement	232-235
Global maximum	64-81
Global stability	130-133

Golden rule	146
Gradient vector	71
Gross substitutability	130-132
Group rationality	207-215
Growth:	
balanced	128
neoclassical model	128-129
von Neumann model	34
Hahn-Negishi process	233-235
Hamiltonian	153
Harrod-Domar model	123
Hawkins-Simon condition	51
Hessian matrix	24
HMMS model of production	148
Homogeneous:	
equation	31
function	36
Hurwicz criterion	189
Hyperplane	75-78
separating	76
Identity matrix	11
Imaginary number	92
Imperfect competition	192
Implicit function:	
rule of differentiation	25-27
theorem	26
Imputation in game	207-209
Imputed value	58
Increasing returns	10
Indecomposability	33
Independence	28
Individual rationality	192-193
Inequality constraints	39-42
Inflation:	
and unemployment	123-125
expectations	231
Inflexion point	79
Information set	220
Initial condition	89-97, 104-106
Inner product	11
Input-coefficient matrix	33

Input-output model:
- dynamic 50-54
- open 33-35
- in relation to linear programming 50-52

Integrand 153
Interior solution 141,153
Intermediate good 33-35
Intersection set 207
Interval 192
Inventory model 141-145
Inverse matrix 11,28
Investment:
- and economic growth 128
- induced 129

Isoquant 27
Iterative method:
- for difference equation 86-88

Kalman-Bucy filtering 178
Karmarkar 42
Keynesian model 123
Koopmans, T.C. 39
Kuhn Tucker:
- conditions 69-71
- Theorem and applications 71-74

Lagrange multiplier 41
- and shadow price 58
- economic interpretation 59-60

Lagrangean function 54
- and saddle point 69-71
- in nonlinear programming 64

Laplace criterion 188
Least squares criterion 53
Leontief model 33-35
Limit of a sequence 19-20
Linear:
- approximation 19
- combination 28
- decision rule 171
- dependence 28
- equations 33-35
- functions 30

inequalities	42
Linear programming	39-48
and duality	54-58
and game theory	60-62
in relation to input-output model	50-54
Local stability analysis	130
Long-run equilibrium	111,123
Lyapunov function	130-131
Lyapunov Theorem	130-132
Mansfield, E.	127
Marginal:	
cost	37
utility	146
Market models	9,28
comparative statics of	28
dynamics of	116
Market shares solution	197
Markov property	175
Markowitz-Tobin theory	73
Mathematical programming	39-82
applications of	49-53,72-74
Matrix concepts	11-17,28-35
Maximum principle	148-154
applications of	152-154
Maximum yield	121
Method of:	
artificial variables	42-45
Lagrange multipliers	54-56
Metzler-Leontief matrix	33-35
Minimax criterion	188
Minimax:	
regret criterion	189
strategy	190
Theorem	200
Minimum variance	73
Minkowski Theorem	78
Minor:	
bordered	80-81
Mixed strategies	190
Mixed difference-differential equation	133

INDEX

Model of:
- advertisement ... 156
- goodwill ... 163
- Walrasian adjustment ... 130

Modulus ... 105
Monetary policy ... 123-125
Monopolistic firm ... 141-144
Move of a game ... 182-185
Multiplier model ... 123
    and acceleration principle ... 125
Muth, John ... 226
Nash bargaining solution ... 198
Nash equilibrium ... 200
National income models ... 129
Necessary conditions ... 39-79
    and sufficiency conditions ... 24, 71
Negative definiteness ... 79-82
    and semi-definiteness ... 80-82
Negotiation set ... 198-200
Neighborhood ... 21-23
Neoclassical optimal growth ... 146-148
Nerlove ... 116
Nonbasic activities (variables) ... 44-46
Noncooperative game ... 200-204
Nonlinear programming ... 64-78
    applications of ... 72-78
Nonnegativity constraints ... 39-46
Nonsingularity ... 28-30
Non-tâtonnement process ... 232-236
Non-Walrasian adjustment ... 234
Nonzero-sum game ... 200-203
Normative rules ... 39
Nucleolus ... 213
Objective function ... 39
Oligopoly ... 192-193
Open-access fishing model ... 221
Open loop solution:
- bilateral monopoly ... 205
- Cournot-Nash ... 205
- Nash bargaining ... 206
- Pareto ... 205

Stackelberg	206
Open set	75
Optimal economic growth	146-148
Optimal portfolio selection	73
Optimality over time	153-170
Optimization:	
dynamic	140-178
static	39-82
stochastic	220-225
Oscillations	92, 105
Payoff	185, 207
function	209
matrix	200-203
Pareto optimality	196, 75-78
Pratt	10
Quadratic:	
form	28-32
programming	69-72
Quasi-competitive solution	196
Quasi-concave:	72
function	72
programming	71-73
Pareto solution	193
particular solution	97, 111
peak-load pricing	49
Perfect:	
competition	141
foresight condition	166-169
Phase plane analysis	133
Phelps-Swan rule	147
Phillips' curve	123-124
Phillips' stabilization policies	124
Pivot rule in simplex	44-46
Portfolio selection problem	73
Positive definiteness	28-32
Predator-prey dynamics	127-128
Price:	
expectations	116
shadow	58
Primal program	54-56
Principle of optimality	171-175

Principle of reciprocity	193-195
Production possibility set	75-78
Production set	76-78
Pure exchange economy	130
Pure strategies	187-191
Radian measure	105
Ramsay rule	147
Rate of growth	128
Rational expectations	220-231
Reaction curves	192-195
Remainder term	21-22
Returns to scale	10
Riccati:	
equation	171-178
Risk attitude	10
risk averse	10
risk neutral	10
risk taker	10
Risk aversion	10, 220-224
Roots:	
characteristic	28-32
complex versus real	92, 105
of quadratic equation	92, 105
Routh-Hurwitz Theorem	111
Saddle point:	
and duality	54-56
and Kuhn-Tucker Theorem	68-72
of optimal trajectory	146-148
Samuelson	147
Samulson model	123-125
Satisficing model	53
Saving ratio	146
Scalar number system	7-8
Schmeidler, D.	213
Schumpeter-Ramsay rule	147
Schur Theorem	97
Second-order conditions	24
in concavity and convexity	71-72
in quasi-concavity and quasi-convexity	69-71
Separating hyperplane	75-78
Separation property	170-172

Set concepts	75, 207-210
Shadow price	49-54
Shapley value	203
Simon, Herbert	53
Simplex algorithm	42-46
Simultaneous equations:	
difference	101
differential	111
Sine function	105
Singular matrix	11-13, 28-30
Sinusoidal function	105
Social welfare function	166
Solow model of growth	128
Solution algorithms	42-46
Stability analysis:	
of general equilibrium	130-132
of Keynesian system	123-125
of market models	28
Stability-instability property	166
Stable equilibrium	92, 116
Stackelberg solution	195
Static analysis	86-90, 140-144
and dynamics	88-105
Steady state	101, 111, 163
Stochastic:	
control	220
difference equation	224-226
differential equation	221-225
equilibrium	220
Strategy:	
maximin	188
minimax	189, 197
mixed	200-203
pure	187
Strong duality	54-58
Substitution Theorem	33-35
Sufficient conditions	79-81
Supply function:	
lagged	116
with price expectations	141-144
Symmetric matrix	25

Tâtonnement process	130
Taylor series expansion	21
Terminal:	
condition	153-155
Test of:	
constraint qualifications (CQ)	69
Threat payoff	204
Trajectory	141, 153-155
Transpose	11-17
Transversality conditions	159-160
Turnpike Theorem	145-147
Two person:	
nonzero-sum game	200-203
zero-sum game	187-190
Types of control	158-166
Unemployment:	
and inflation	123-124
and monetary policy	124-125
van der Pol equation	130
von Neumann-Morgenstern	207
Value of a game	208
and the core	208-210
Vector:	
characteristic	28-30
linearly dependent	28
linearly independent	28
unit	198
Vector maximum problem	196
Vertices:	
in game tree	182
in LP models	42-44
Von Neumann-Morgenstern:	
solution in game	207-208
utility function	208-210
Wage-price frontier	158-170
Wald, Abraham	188
Walras law	130, 234
Weak duality	54-58
Weierstrass Theorem	78
Zero-sum game	187
and bilinear payoff	200-204